An Introduction to
Kalman Filtering
with MATLAB Examples

Synthesis Lectures on Signal Processing

Editor
José Moura, *Carnegie Mello University*

Synthesis Lectures in Signal Processing will publish 50- to 100-page books on topics of interest to signal processing engineers and researchers. The Lectures exploit in detail a focused topic. They can be at different levels of exposition—from a basic introductory tutorial to an advanced monograph—depending on the subject and the goals of the author. Over time, the Lectures will provide a comprehensive treatment of signal processing. Because of its format, the Lectures will also provide current coverage of signal processing, and existing Lectures will be updated by authors when justified.

Lectures in Signal Processing are open to all relevant areas in signal processing. They will cover theory and theoretical methods, algorithms, performance analysis, and applications. Some Lectures will provide a new look at a well established area or problem, while others will venture into a brand new topic in signal processing. By careful reviewing the manuscripts we will strive for quality both in the Lectures' contents and exposition.

An Introduction to Kalman Filtering with MATLAB Examples

Narayan Kovvali, Mahesh Banavar, and Andreas Spanias

ISBN: 978-3-031-01408-6   paperback
ISBN: 978-3-031-02536-5   ebook

DOI 10.1007/978-3-031-02536-5

A Publication in the Springer series
SYNTHESIS LECTURES ON SIGNAL PROCESSING

Lecture #12
Series Editor: José Moura, *Carnegie Mello University*
Series ISSN
Synthesis Lectures on Signal Processing
Print 1932-1236   Electronic 1932-1694

An Introduction to
Kalman Filtering
with MATLAB Examples

Narayan Kovvali, Mahesh Banavar, and Andreas Spanias

SenSIP Center, Arizona State University

SYNTHESIS LECTURES ON SIGNAL PROCESSING #12

ABSTRACT

The Kalman filter is the Bayesian optimum solution to the problem of sequentially estimating the states of a dynamical system in which the state evolution and measurement processes are both linear and Gaussian. Given the ubiquity of such systems, the Kalman filter finds use in a variety of applications, e.g., target tracking, guidance and navigation, and communications systems. The purpose of this book is to present a brief introduction to Kalman filtering. The theoretical framework of the Kalman filter is first presented, followed by examples showing its use in practical applications. Extensions of the method to nonlinear problems and distributed applications are discussed. A software implementation of the algorithm in the MATLAB programming language is provided, as well as MATLAB code for several example applications discussed in the manuscript.

KEYWORDS

dynamical system, parameter estimation, tracking, state space model, sequential Bayesian estimation, linearity, Gaussian noise, Kalman filter

Contents

Acknowledgments

The work in this book was supported in part by the SenSIP Center, Arizona State University.

Narayan Kovvali, Mahesh Banavar, and Andreas Spanias
September 2013

CHAPTER 1

Introduction

Statistical estimation is the process of determining the values of certain parameters or signals from empirical (measured or collected) data which is typically noisy and random in nature [1, 2]. Statistical estimation has application in a multitude of areas. For example, in the area of consumer electronics, estimation techniques are used for mobile wireless communications, intelligent voice and gesture recognition, multimedia enhancement and classification, GPS navigation, and much more. In defense and security related fields, applications include target tracking, guidance and navigation systems, and threat detection. Statistical estimation methods also play a vital role in health monitoring and medical diagnosis problems.

The flow graph of an estimation problem is shown in Figure 1.1. In the estimation problem, the task is to estimate an unobservable phenomenon of interest (represented by a set of parameters) using observed data. When the parameters vary over time, the estimate may be iteratively updated using continuously obtained data, as shown in Figure 1.2. This type of estimation, termed sequential estimation, involves computing an initial estimate and then iteratively updating the estimate based on the most recent data.

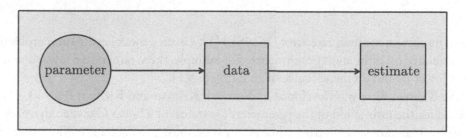

Figure 1.1: In the estimation problem, the task is to estimate an unobservable phenomenon of interest (represented by a set of parameters) using observed data.

A large number of statistical estimation algorithms exist, ranging from point estimators such as maximum-likelihood (ML) and maximum a posteriori (MAP) estimators which compute the single best parameter that maximizes the likelihood of the observed data or a posteriori parameter probability, to Bayesian methods which compute the full posterior probability distribution of the parameters. One class of estimation, known as linear estimation, computes the parameter estimate as a simple linear function of the data. Estimation algorithms are designed to satisfy various optimality criteria, such as consistency, efficiency, unbiasedness, minimum vari-

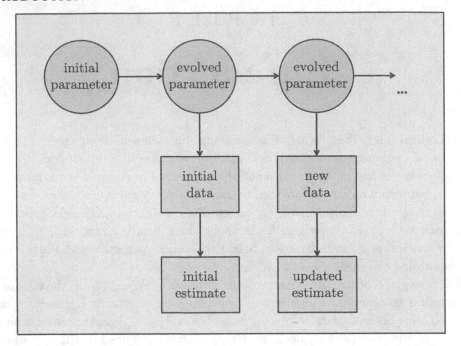

Figure 1.2: In sequential estimation, time-varying parameters are estimated by computing an initial estimate and then iteratively updating the estimate based on the most recent data.

ance, and minimum mean square error (MMSE) [1]. Given a model estimation problem, bounds can be calculated on estimation performance. For example, the Cramér-Rao lower bound (CRLB) defines a lower bound on the variance of an estimator [1].

The Kalman filter was developed by Rudolph Kalman and Richard Bucy [3, 4] in the 60s and is used for the estimation of the parameters (or states) of a linear Gaussian dynamical system. Specifically, in a state space setting, the system state must propagate according to a linear evolution model with Gaussian process noise and the data (measurements) must be linearly related to the state with Gaussian noise. Many systems encountered in real-world applications are well-characterized by this model. The Kalman filter is a popular choice for estimating the parameters of dynamical systems for several reasons, including:

- The Kalman filter is the Bayesian optimum estimator for sequentially estimating the states of a linear Gaussian dynamical system;

- The Kalman filtering algorithm has low computational complexity and can be implemented in DSP hardware for realtime applications;

- Variations and extensions of the Kalman filter are readily available for nonlinear, distributed, and non-Gaussian problems, such as the extended Kalman filter, the unscented Kalman filter, the decentralized Kalman filter, and the particle filter [5].

The Kalman filter finds application in a variety of problems, for example, target tracking [6] and sensor drift correction and inertial navigation [7–9].

The rest of this book is organized as follows. Chapter 2 introduces the statistical estimation problem and reviews important estimation approaches, such as maximum-likelihood and Bayesian estimators. Chapter 3 describes the analytical framework of the Kalman filter. Chapter 4 discusses extensions of Kalman filtering for nonlinear problems and distributed applications. In each chapter, several examples are presented to illustrate the methods and algorithms. Figures are used in the form of block diagrams, system descriptions, and plots throughout the manuscript to illustrate the concepts. MATLAB programs are also provided for most examples. Chapter 5 summarizes the topics covered and provides some references for further reading.

CHAPTER 2

The Estimation Problem

2.1 BACKGROUND

The problem of estimating nonobservable phenomena of interest (termed *parameters*) using observed data (e.g., sensor measurements) is a ubiquitous one. Estimation tasks arise in a wide variety of real-world applications, such as medical diagnosis using imaging and bio-signals, weather forecasting based on temperature, pressure, and wind speed readings, and the tracking of target position and velocity using radar or sonar imagery.

A general parameter estimation problem [1, 10, 11] can be stated as follows. Let $\mathbf{x}$ be a $D \times 1$ vector that denotes the parameters of interest and $\mathbf{y}$ a $M \times 1$ vector denoting observed data. The data $\mathbf{y}$ carries information about parameters $\mathbf{x}$, quantified using a *measurement model* that explicitly relates $\mathbf{x}$ and $\mathbf{y}$. The measurement model could be deterministic (i.e., it involves no random components), for example,

$$\mathbf{y} = \mathbf{h}(\mathbf{x}), \tag{2.1}$$

where $\mathbf{h} : \mathbb{R}^D \mapsto \mathbb{R}^M$ is a known function (we assume, without loss of generality, that the parameters and data are real-valued). It could also be probabilistic, with a stochastic model used to describe the relationship between $\mathbf{x}$ and $\mathbf{y}$:

$$\mathbf{y} = \mathbf{h}(\mathbf{x}) + \mathbf{w}, \tag{2.2}$$

where $\mathbf{w}$ is a (random) $M \times 1$ measurement noise vector, or, more generally,

$$\mathbf{y} \sim p(\mathbf{y}|\mathbf{x}), \tag{2.3}$$

which directly specifies the conditional probability density function (pdf) of $\mathbf{y}$ given $\mathbf{x}$. The estimation problem is then to determine the parameter vector $\mathbf{x}$ using the observed data $\mathbf{y}$, and an estimation algorithm is employed to compute the estimate of $\mathbf{x}$, denoted as $\hat{\mathbf{x}}$. The estimate $\hat{\mathbf{x}}(\mathbf{y})$ depends on the data $\mathbf{y}$. For the models in (2.1) and (2.2), this amounts to some sort of an inversion of the function $\mathbf{h}$:

$$\hat{\mathbf{x}}(\mathbf{y}) \approx \mathbf{h}^{-1}(\mathbf{y}). \tag{2.4}$$

For (2.3), a popular estimator for $\mathbf{x}$ is the *maximum-likelihood (ML) estimate* [1, 10, 11]:

$$\hat{\mathbf{x}}^{\text{ML}}(\mathbf{y}) \triangleq \underset{\mathbf{x}}{\operatorname{argmax}} \ p(\mathbf{y}|\mathbf{x}). \tag{2.5}$$

The ML estimator is asymptotically optimal or *consistent*: $\hat{\mathbf{x}}^{\mathrm{ML}}$ converges in probability to $\mathbf{x}$ as the data set size $N \to \infty$. An estimate $\hat{\mathbf{x}}$ is said to be *unbiased* if

$$E[\hat{\mathbf{x}}(\mathbf{y})] = \int \hat{\mathbf{x}}(\mathbf{y})\, p(\mathbf{y}|\mathbf{x})\, d\mathbf{y} = \mathbf{x}, \tag{2.6}$$

where $E[\cdot]$ denotes expectation. Figure 2.1 shows the basic components of the general estimation problem.

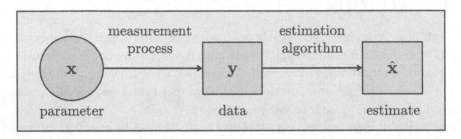

Figure 2.1: The general estimation problem is to determine a parameter vector of interest $\mathbf{x}$ using observed data $\mathbf{y}$, and an estimation algorithm is employed to compute the estimate $\hat{\mathbf{x}}$.

In many applications, we are interested in estimating parameters that are not static but rather dynamic or varying over time, with the dynamics governed by an evolution law or model. A large number of real-world dynamical systems characterized by stationary and non-stationary random processes fall into this category. For the purpose of estimation, data about the time-varying parameters is typically collected at regular intervals of time. In this work we will only consider discrete-time dynamical systems.

2.1.1   EXAMPLE: MAXIMUM-LIKELIHOOD ESTIMATION IN GAUSSIAN NOISE

Consider the problem of estimating parameters given data corrupted by additive Gaussian noise. Specifically, we are interested in estimating parameters $\mathbf{x}$ from noisy data $\mathbf{y}$ of the form

$$\mathbf{y} = \mathbf{h}(\mathbf{x}) + \mathbf{w}, \tag{2.7}$$

where the function $\mathbf{h}: \mathbb{R}^D \mapsto \mathbb{R}^M$ models the relationship between the $D \times 1$ parameter vector $\mathbf{x}$ and $M \times 1$ data vector $\mathbf{y}$, and $\mathbf{w}$ is a $M \times 1$ Gaussian random measurement noise vector with zero mean and covariance matrix $\mathbf{R} = \sigma_y^2 \mathbf{I}$. The ML estimator $\hat{\mathbf{x}}^{\mathrm{ML}}(\mathbf{y})$ for the parameters $\mathbf{x}$ is obtained as follows.

The likelihood function for the parameters $\mathbf{x}$ is given by

$$p(\mathbf{y}|\mathbf{x}) = \frac{1}{(2\pi\sigma_y^2)^{M/2}} \exp\left(-\frac{1}{2\sigma_y^2} \|\mathbf{y} - \mathbf{h}(\mathbf{x})\|^2\right), \tag{2.8}$$

where $\|\cdot\|$ denotes the 2-norm, and the ML estimator is

$$
\begin{aligned}
\hat{\mathbf{x}}^{\mathrm{ML}}(\mathbf{y}) &= \underset{\mathbf{x}}{\operatorname{argmax}} \log p(\mathbf{y}|\mathbf{x}), \\
&= \underset{\mathbf{x}}{\operatorname{argmax}} \left[-\frac{M}{2} \log 2\pi - \frac{M}{2} \log \sigma_y^2 - \frac{1}{2\sigma_y^2} \|\mathbf{y} - \mathbf{h}(\mathbf{x})\|^2 \right], \\
&= \underset{\mathbf{x}}{\operatorname{argmin}} \|\mathbf{y} - \mathbf{h}(\mathbf{x})\|^2,
\end{aligned}
\tag{2.9}
$$

which is the well-known *least squares* solution. In particular, when the model is linear, i.e.,

$$
\mathbf{y} = \mathbf{H}\mathbf{x} + \mathbf{w},
\tag{2.10}
$$

with $\mathbf{H}$ a $M \times D$ matrix, then the ML estimate of the parameters $\mathbf{x}$ is given by

$$
\hat{\mathbf{x}}^{\mathrm{ML}}(\mathbf{y}) = \underset{\mathbf{x}}{\operatorname{argmin}} \|\mathbf{y} - \mathbf{H}\mathbf{x}\|^2 = (\mathbf{H}^T \mathbf{H})^{-1} \mathbf{H}^T \mathbf{y}.
\tag{2.11}
$$

2.2 LINEAR ESTIMATION

Before proceeding to general probabilistic estimation, in this section we briefly discuss the frequently encountered setting of *linear estimation*. For our discussion, we pick the specific context of estimating a sampled discrete-time signal from noisy measurements. Let x_n denote the signal of interest at time step $n \in \mathbb{Z}^+$ and let y_n be the corresponding noisy measured signal. The goal is to design a linear estimator for x_n of the form

$$
\hat{x}_n = \sum_{i=0}^{M-1} a_i y_{n-i},
\tag{2.12}
$$

where $a_0, a_1, \ldots, a_{M-1}$ are filter coefficients, such that the error $x_n - \hat{x}_n$ is minimized. In matrix notation,

$$
\hat{x}_n = \begin{bmatrix} a_0 & a_1 & \ldots & a_{M-1} \end{bmatrix} \begin{bmatrix} y_n \\ y_{n-1} \\ \vdots \\ y_{n-M+1} \end{bmatrix}.
\tag{2.13}
$$

The mean square error is

$$
\begin{aligned}
E[(x_n - \hat{x}_n)^2] &= E[x_n^2] - 2E[x_n \hat{x}_n] + E[\hat{x}_n^2] \\
&= E[x_n^2] - 2E\left[x_n \sum_{i=0}^{M-1} a_i y_{n-i} \right] + E\left[\left(\sum_{i=0}^{M-1} a_i y_{n-i} \right)^2 \right],
\end{aligned}
\tag{2.14}
$$

and differentiating with respect to a_i gives

$$\frac{\partial}{\partial a_i} E[(x_n - \hat{x}_n)^2] = -2E[x_n\, y_{n-i}] + 2E\left[\left(\sum_{j=0}^{M-1} a_j\, y_{n-j}\right) y_{n-i}\right]$$

$$= -2E[x_n\, y_{n-i}] + 2\sum_{j=0}^{M-1} a_j\, E[y_{n-j}\, y_{n-i}], \qquad (2.15)$$

for $i = 0, \ldots, M - 1$. Equating this to zero, we see that the desired filter coefficients satisfy

$$\sum_{j=0}^{M-1} a_j\, E[y_{n-j}\, y_{n-i}] = E[x_n\, y_{n-i}], \quad i = 0, \ldots, M - 1. \qquad (2.16)$$

When the signals y_n and x_n are *wide-sense stationary* (i.e., their statistics up to the second-order are independent of time), with autocorrelation and cross-correlation sequences given respectively by

$$r_i^{yy} \triangleq E[y_n\, y_{n-i}], \qquad (2.17a)$$
$$r_i^{xy} \triangleq E[x_n\, y_{n-i}], \qquad (2.17b)$$

Eq. (2.16) becomes

$$\sum_{j=0}^{M-1} a_j\, r_{i-j}^{yy} = r_i^{xy}, \quad i = 0, \ldots, M - 1, \qquad (2.18)$$

which are the *Wiener-Hopf equations* [12]. Note that the autocorrelation and cross-correlation sequences obey symmetry properties: $r_i^{yy} = r_{-i}^{yy}$ and $r_i^{xy} = r_{-i}^{yx}$. In matrix notation, (2.18) can be written as

$$\begin{bmatrix} r_0^{yy} & r_1^{yy} & \cdots & r_{M-1}^{yy} \\ r_1^{yy} & r_0^{yy} & \cdots & r_{M-2}^{yy} \\ \vdots & \vdots & \ddots & \vdots \\ r_{M-1}^{yy} & r_{M-2}^{yy} & \cdots & r_0^{yy} \end{bmatrix} \begin{bmatrix} a_0 \\ a_1 \\ \vdots \\ a_{M-1} \end{bmatrix} = \begin{bmatrix} r_0^{xy} \\ r_1^{xy} \\ \vdots \\ r_{M-1}^{xy} \end{bmatrix}, \qquad (2.19)$$

or

$$\mathbf{R}^{yy}\, \mathbf{a} = \mathbf{r}^{xy}. \qquad (2.20)$$

The *Wiener filter* coefficients are given by

$$\mathbf{a} = (\mathbf{R}^{yy})^{-1}\, \mathbf{r}^{xy}. \qquad (2.21)$$

The matrix $\mathbf{R}^{yy}$ is symmetric and positive-semidefinite, and a unique solution almost always exists for (2.21). Since the signals involved are stationary, the filter is time-invariant. The linear estimate for x_n is then computed using (2.13) as

$$\hat{x}_n = \mathbf{a}^T\, \mathbf{y}_n, \qquad (2.22)$$

where $\mathbf{a} = [a_0 \; a_1 \; \ldots \; a_{M-1}]^T$ obtained in (2.21), and $\mathbf{y}_n = [y_n \; y_{n-1} \; \ldots \; y_{n-M+1}]^T$ is the $M \times 1$ measurement vector at time step n.

2.3 THE BAYESIAN APPROACH TO PARAMETER ESTIMATION

In the Bayesian approach [1, 10, 11, 13] to estimation, the parameters of interest are probabilistically estimated by combining two pieces of information: (a) belief about the parameters based on the likelihood of the observed data as stipulated by a probabilistic measurement model, and (b) a priori knowledge about the parameters quantified probabilistically by a prior pdf. Specifically, in Bayesian inference the posterior pdf $p(\mathbf{x}|\mathbf{y})$ over the parameters $\mathbf{x}$ given the data $\mathbf{y}$ is computed by combining the likelihood $p(\mathbf{y}|\mathbf{x})$ and prior $p(\mathbf{x})$ using Bayes' theorem [11]:

$$p(\mathbf{x}|\mathbf{y}) = \frac{p(\mathbf{y}|\mathbf{x})\,p(\mathbf{x})}{p(\mathbf{y})} = \frac{p(\mathbf{y}|\mathbf{x})\,p(\mathbf{x})}{\int p(\mathbf{y}|\mathbf{x})\,p(\mathbf{x})\,d\mathbf{x}} \propto p(\mathbf{y}|\mathbf{x})\,p(\mathbf{x}). \tag{2.23}$$

Eq. (2.23) prescribes how a priori belief $p(\mathbf{x})$ about the parameter $\mathbf{x}$ is updated to the a posteriori belief $p(\mathbf{x}|\mathbf{y})$ in light of the observed data $\mathbf{y}$. The *Bayes estimate* is typically defined as the mean of the posterior pdf:

$$\hat{\mathbf{x}}^{B}(\mathbf{y}) \triangleq E[\mathbf{x}|\mathbf{y}] = \int \mathbf{x}\,p(\mathbf{x}|\mathbf{y})\,d\mathbf{x}. \tag{2.24}$$

The Bayes estimator in (2.24) is optimum in the sense that it minimizes the *mean square error (MSE)* Bayes risk [1]

$$E[(\mathbf{x} - \hat{\mathbf{x}}^{B}(\mathbf{y}))^{T}(\mathbf{x} - \hat{\mathbf{x}}^{B}(\mathbf{y}))] = \iint (\mathbf{x} - \hat{\mathbf{x}}^{B}(\mathbf{y}))^{T}(\mathbf{x} - \hat{\mathbf{x}}^{B}(\mathbf{y}))\,p(\mathbf{x},\mathbf{y})\,d\mathbf{x}\,d\mathbf{y}. \tag{2.25}$$

In other words

$$\hat{\mathbf{x}}^{B}(\mathbf{y}) = \underset{\hat{\mathbf{x}}(\mathbf{y})}{\operatorname{argmin}}\; E[(\mathbf{x} - \hat{\mathbf{x}}(\mathbf{y}))^{T}(\mathbf{x} - \hat{\mathbf{x}}(\mathbf{y}))]. \tag{2.26}$$

Additionally, the Bayes estimator is asymptotically unbiased.

As seen in (2.23) and (2.24), Bayesian estimation utilizes Bayes' theorem to compute the desired parameter estimate. Bayes' theorem in its current form was developed as a generalization of the work of Bayes [14]. With the advent of powerful computers, Bayesian estimation is applied today in many real-world problems with complex statistical models using Monte Carlo techniques [15, 16].

2.3.1 EXAMPLE: ESTIMATING THE BIAS OF A COIN

In a coin flipping experiment, a coin is tossed N times and the outcome 'heads' is observed r times. Based on this data, we wish to estimate the probability of heads p_H for this coin. Following the

Bayesian approach, we start by defining a prior pdf over p_H, taken here to be the standard uniform distribution:

$$p(p_H) = \begin{cases} 1, & \text{for } 0 \le p_H \le 1, \\ 0, & \text{otherwise.} \end{cases} \tag{2.27}$$

Given p_H, the probability of observing heads r times in N coin tosses is Binomial and given by

$$p(r|N, p_H) = \binom{N}{r} p_H^r (1 - p_H)^{N-r}, \quad \text{for } r = 0, \ldots, N. \tag{2.28}$$

Combining the prior in (2.27) with the likelihood in (2.28) using Bayes' theorem, we obtain the posterior pdf of p_H as the Beta distribution:

$$p(p_H|r, N) \propto p(r|N, p_H)\, p(p_H) = \begin{cases} \frac{1}{B(r+1, N-r+1)} p_H^r (1 - p_H)^{N-r}, & \text{for } 0 \le p_H \le 1, \\ 0, & \text{otherwise,} \end{cases} \tag{2.29}$$

where $B(\cdot, \cdot)$ denotes the Beta function. The Bayes estimate of p_H is then the mean of the Beta posterior pdf, $\text{Beta}(p_H; r + 1, N - r + 1)$, and is given by [2]

$$\hat{p}_H^B = E[p_H|r, N] = \int_0^1 p_H\, \text{Beta}(p_H; r + 1, N - r + 1)\, dp_H = \frac{r + 1}{N + 2}. \tag{2.30}$$

It should be mentioned here that the ML estimate of p_H is

$$\hat{p}_H^{\text{ML}} = \underset{p_H}{\operatorname{argmax}}\ p(r|N, p_H) = \underset{p_H}{\operatorname{argmax}}\ \binom{N}{r} p_H^r (1 - p_H)^{N-r} = \frac{r}{N}. \tag{2.31}$$

Figure 2.2f shows plots of the Beta posterior pdf for various r and N. Observe that the variance of the posterior pdfs decreases with increasing data. The ML estimation method simply yields a point estimate, unlike the Bayesian estimation approach which provides the full posterior pdf that can be used to assess confidence in the estimate.

2.4   SEQUENTIAL BAYESIAN ESTIMATION

In dynamical systems with parameters evolving over time, a convenient framework for representation is provided by *state space models* of the form

$$\mathbf{x}_n \sim p(\mathbf{x}_n|\mathbf{x}_{n-1}), \tag{2.32a}$$
$$\mathbf{y}_n \sim p(\mathbf{y}_n|\mathbf{x}_n), \tag{2.32b}$$

where $\mathbf{x}_n$ is the $D \times 1$ state vector (parameters) at time step $n \in \mathbb{Z}^+$ and $\mathbf{y}_n$ is the corresponding $M \times 1$ measurement vector (data). The state vector $\mathbf{x}_n$ might, for example, represent the position and velocity of a moving target at time step n that need to be estimated, with the measurement

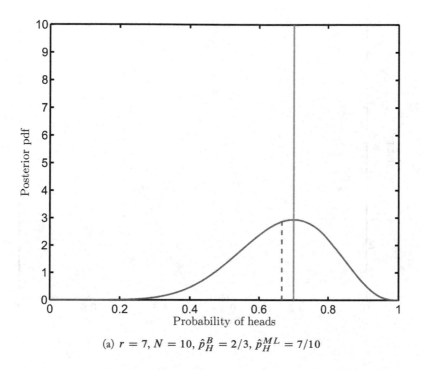

(a) $r = 7$, $N = 10$, $\hat{p}_H^B = 2/3$, $\hat{p}_H^{ML} = 7/10$

Figure 2.2a: Bayesian estimation of the probability of heads p_H for a coin, when the outcome 'heads' is observed to occur r times in N coin tosses. The plots show the posterior pdf $p(p_H|r, N)$, which, under a uniform prior, is the Beta distribution $\text{Beta}(r + 1, N - r + 1)$. The Bayes and ML estimates of p_H are $r + 1/N + 2$ and r/N, respectively.

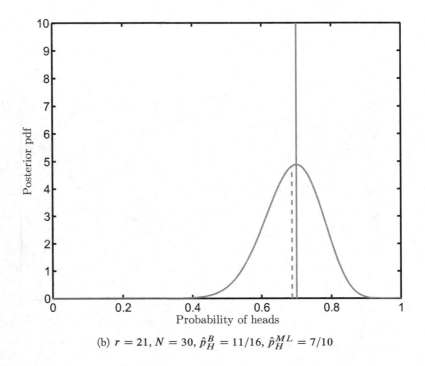

(b) $r = 21$, $N = 30$, $\hat{p}_H^B = 11/16$, $\hat{p}_H^{ML} = 7/10$

Figure 2.2b: Bayesian estimation of the probability of heads p_H for a coin, when the outcome 'heads' is observed to occur r times in N coin tosses. The plots show the posterior pdf $p(p_H|r, N)$, which, under a uniform prior, is the Beta distribution Beta($r + 1, N - r + 1$). The Bayes and ML estimates of p_H are $r + 1/N + 2$ and r/N, respectively.

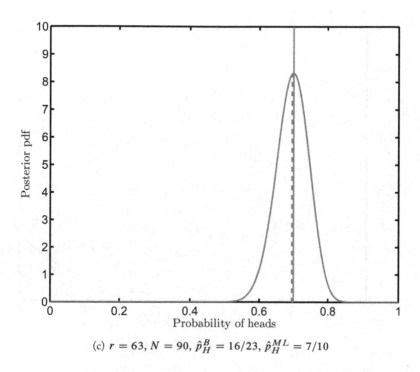

(c) $r = 63$, $N = 90$, $\hat{p}_H^B = 16/23$, $\hat{p}_H^{ML} = 7/10$

Figure 2.2c: Bayesian estimation of the probability of heads p_H for a coin, when the outcome 'heads' is observed to occur r times in N coin tosses. The plots show the posterior pdf $p(p_H|r, N)$, which, under a uniform prior, is the Beta distribution $\text{Beta}(r + 1, N - r + 1)$. The Bayes and ML estimates of p_H are $r + 1/N + 2$ and r/N, respectively.

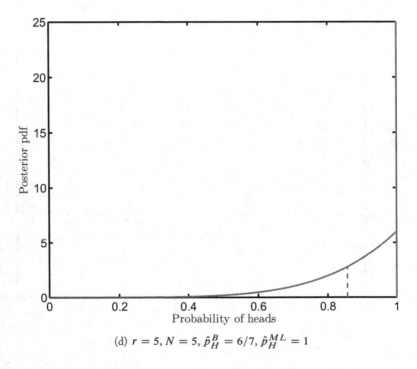

(d) $r = 5, N = 5, \hat{p}_H^B = 6/7, \hat{p}_H^{ML} = 1$

Figure 2.2d: Bayesian estimation of the probability of heads p_H for a coin, when the outcome 'heads' is observed to occur r times in N coin tosses. The plots show the posterior pdf $p(p_H|r, N)$, which, under a uniform prior, is the Beta distribution Beta$(r + 1, N - r + 1)$. The Bayes and ML estimates of p_H are $r + 1/N + 2$ and r/N, respectively.

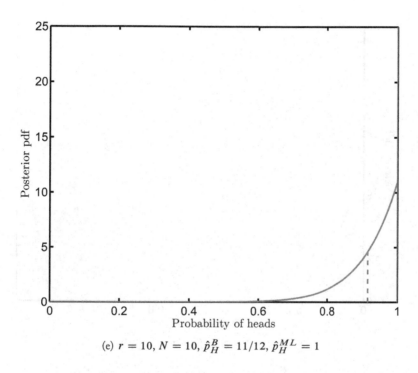

(e) $r = 10$, $N = 10$, $\hat{p}_H^B = 11/12$, $\hat{p}_H^{ML} = 1$

Figure 2.2e: Bayesian estimation of the probability of heads p_H for a coin, when the outcome 'heads' is observed to occur r times in N coin tosses. The plots show the posterior pdf $p(p_H | r, N)$, which, under a uniform prior, is the Beta distribution $\text{Beta}(r + 1, N - r + 1)$. The Bayes and ML estimates of p_H are $r + 1/N + 2$ and r/N, respectively.

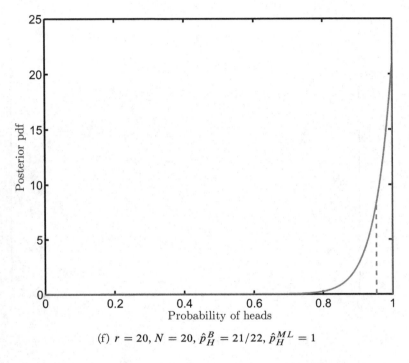

(f) $r = 20, N = 20, \hat{p}_H^B = 21/22, \hat{p}_H^{ML} = 1$

Figure 2.2f: Bayesian estimation of the probability of heads p_H for a coin, when the outcome 'heads' is observed to occur r times in N coin tosses. The plots show the posterior pdf $p(p_H|r, N)$, which, under a uniform prior, is the Beta distribution Beta$(r + 1, N - r + 1)$. The Bayes and ML estimates of p_H are $r + 1/N + 2$ and r/N, respectively.

vector $\mathbf{y}_n$ comprised of range and range-rate information collected at time step n by a radar system. Eq. (2.32a) specifies a first-order *Markov* [2] state evolution law for the time-varying state, in which the state $\mathbf{x}_n$ at any time step n is independent of the states $\mathbf{x}_{n-2}, \mathbf{x}_{n-3}, \ldots, \mathbf{x}_0$ at the past time steps $n-2, n-3, \ldots, 0$ given the state $\mathbf{x}_{n-1}$ at the immediate previous time step $n-1$. Eq. (2.32b) describes the relationship between the time-varying state and the measurements, in which the measurement $\mathbf{y}_n$ at any time step n is conditionally independent of all other states given the state $\mathbf{x}_n$ at that time step n. Note that both state and measurement models are probabilistic. The initial state distribution is given by the pdf $p(\mathbf{x}_0)$. Figure 2.3 shows graphically the state space model for a dynamical system.

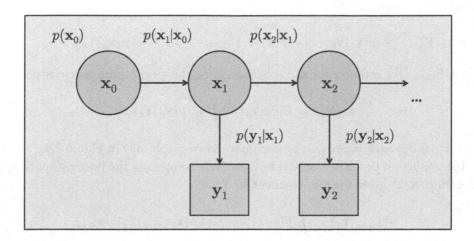

Figure 2.3: State space model for a dynamical system. The time-varying state evolves according to a first-order Markov model, in which the state $\mathbf{x}_n$ at any time step n is independent of the states $\mathbf{x}_{n-2}, \mathbf{x}_{n-3}, \ldots, \mathbf{x}_0$ at the past time steps $n-2, n-3, \ldots, 0$ given the state $\mathbf{x}_{n-1}$ at the immediate previous time step $n-1$. The measurement $\mathbf{y}_n$ at any time step n is conditionally independent of all other states given the state $\mathbf{x}_n$ at that time step n. The initial state distribution is given by the pdf $p(\mathbf{x}_0)$.

In the state space setting, two key tasks of interest are: (a) *filtering*, and (b) *prediction*. The filtering problem is to estimate the state $\mathbf{x}_n$ at time step n given the set $\mathbf{Y}_n \triangleq \{\mathbf{y}_1, \mathbf{y}_2, \ldots, \mathbf{y}_n\}$ of measurements up to time step n. The prediction problem is to estimate the state(s) $\mathbf{x}_{n+l}$ at time step $n+l, l > 0$, given the set $\mathbf{Y}_n$ of measurements up to time step n.

The Bayesian optimum solution to the filtering problem sequentially or recursively computes the posterior pdf $p(\mathbf{x}_n|\mathbf{Y}_n)$ over the state $\mathbf{x}_n$ given the measurement set $\mathbf{Y}_n$ in two steps. In the first step, known as the prediction step, the Markov state evolution model $p(\mathbf{x}_n|\mathbf{x}_{n-1})$ is used with the posterior pdf $p(\mathbf{x}_{n-1}|\mathbf{Y}_{n-1})$ at the previous time step $n-1$ to obtain the pdf $p(\mathbf{x}_n|\mathbf{Y}_{n-1})$ over the state $\mathbf{x}_n$ given the measurement set $\mathbf{Y}_{n-1}$. In the second step, known as the update step, the posterior pdf $p(\mathbf{x}_n|\mathbf{Y}_n)$ is computed by combining the measurement likelihood $p(\mathbf{y}_n|\mathbf{x}_n)$ and prediction $p(\mathbf{x}_n|\mathbf{Y}_{n-1})$ using Bayes' theorem. Mathematically,

$$\text{Predict step:} \qquad p(\mathbf{x}_n|\mathbf{Y}_{n-1}) = \int p(\mathbf{x}_n|\mathbf{x}_{n-1})\, p(\mathbf{x}_{n-1}|\mathbf{Y}_{n-1})\, d\mathbf{x}_{n-1}, \qquad (2.33a)$$

$$\text{Update step:} \qquad p(\mathbf{x}_n|\mathbf{Y}_n) \propto p(\mathbf{y}_n|\mathbf{x}_n)\, p(\mathbf{x}_n|\mathbf{Y}_{n-1}), \qquad (2.33b)$$

for $n = 1, 2, \ldots$, with $p(\mathbf{x}_0|\mathbf{Y}_0) \triangleq p(\mathbf{x}_0)$ used to initialize the iteration. Observe, by comparing with (2.23), that at each time step here the prediction $p(\mathbf{x}_n|\mathbf{Y}_{n-1})$ plays the role of the prior pdf. The Bayesian filtering iteration of Eq. (2.33) is a recursive computation of the posterior pdf $p(\mathbf{x}_n|\mathbf{Y}_n)$ at each time step n:

$$p(\mathbf{x}_0|\mathbf{Y}_0) \overset{\text{predict}}{\longrightarrow} p(\mathbf{x}_1|\mathbf{Y}_0) \overset{\text{update}}{\longrightarrow} p(\mathbf{x}_1|\mathbf{Y}_1) \overset{\text{predict}}{\longrightarrow} \ldots\ldots \overset{\text{update}}{\longrightarrow} p(\mathbf{x}_n|\mathbf{Y}_n) \overset{\text{predict}}{\longrightarrow} \ldots \qquad (2.34)$$

for $n = 1, 2, \ldots$. The estimate for $\mathbf{x}_n$ is obtained using the mean of the posterior pdf:

$$\hat{\mathbf{x}}_n(\mathbf{Y}_n) = E[\mathbf{x}_n|\mathbf{Y}_n] = \int \mathbf{x}_n\, p(\mathbf{x}_n|\mathbf{Y}_n)\, d\mathbf{x}_n. \qquad (2.35)$$

The sequential Bayesian estimation procedure is shown graphically in Figure 2.4.

The prediction problem is solved by recursively computing the posterior pdfs $p(\mathbf{x}_{n+l}|\mathbf{Y}_n)$ over the states $\mathbf{x}_{n+l}$ given the measurement set $\mathbf{Y}_n$ as

$$p(\mathbf{x}_{n+l}|\mathbf{Y}_n) = \int p(\mathbf{x}_{n+l}|\mathbf{x}_{n+l-1})\, p(\mathbf{x}_{n+l-1}|\mathbf{Y}_n)\, d\mathbf{x}_{n+l-1}, \qquad (2.36)$$

for $l = 1, 2, \ldots$, and the estimates for $\mathbf{x}_{n+l}$ are again the means of the posterior pdfs.

2.4.1 EXAMPLE: THE 1-D KALMAN FILTER

Consider a one-dimensional (1-D) linear Gaussian state space model

$$x_n = F_n\, x_{n-1} + v_n, \qquad (2.37a)$$
$$y_n = H_n\, x_n + w_n, \qquad (2.37b)$$

where x_n is a scalar state at time step n, y_n is the corresponding scalar measurement, F_n is a linear state-transition parameter, v_n is Gaussian random state noise with zero mean and variance Q_n, H_n is a linear measurement parameter, and w_n is Gaussian random measurement noise with zero mean and variance R_n. The initial state distribution $p(x_0)$ is Gaussian. The task is to sequentially estimate the state x_n using the set of measurements $Y_n = \{y_1, y_2, \ldots, y_n\}$.

We follow the sequential Bayesian estimation procedure described earlier. Suppose that at time step $n-1$ the posterior pdf $p(x_{n-1}|Y_{n-1})$ is Gaussian:

$$p(x_{n-1}|Y_{n-1}) \equiv \mathcal{N}(x_{n-1}; m_{n-1|n-1}, P_{n-1|n-1}), \qquad (2.38)$$

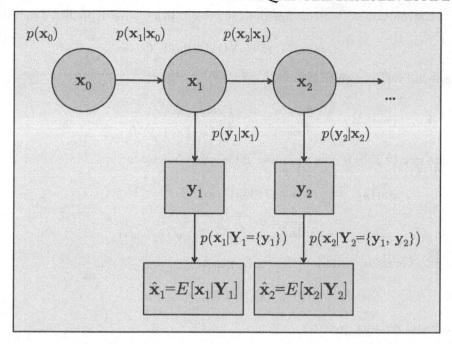

Figure 2.4: Sequential Bayesian estimation recursively computes the posterior pdf $p(\mathbf{x}_n|\mathbf{Y}_n)$ over the state $\mathbf{x}_n$ given the measurement set $\mathbf{Y}_n = \{\mathbf{y}_1, \mathbf{y}_2, \ldots, \mathbf{y}_n\}$. The estimate for $\mathbf{x}_n$ is obtained using the mean of the posterior pdf: $\hat{\mathbf{x}}_n(\mathbf{Y}_n) = E[\mathbf{x}_n|\mathbf{Y}_n]$.

where the notation $\mathcal{N}(\cdot\,; m, P)$ is used to denote a Gaussian pdf with mean m and variance P. Thus in (2.38) $m_{n-1|n-1}$ and $P_{n-1|n-1}$ denote, respectively, the Gaussian posterior mean and variance at time step $n-1$. The subscript notation '$n-1|n-1$' is used to indicate association with the (Gaussian) pdf of the state x_{n-1} at time step $n-1$ computed using the measurements up to time step $n-1$. From the model (2.37), the conditional pdfs $p(x_n|x_{n-1})$ and $p(y_n|x_n)$ of the state and measurement, respectively, are given by the Gaussian pdfs

$$
\begin{aligned}
p(x_n|x_{n-1}) &\equiv \mathcal{N}(x_n; F_n\, x_{n-1}, Q_n), &(2.39\text{a})\\
p(y_n|x_n) &\equiv \mathcal{N}(y_n; H_n\, x_n, R_n). &(2.39\text{b})
\end{aligned}
$$

The prediction step (2.33a) therefore becomes

$$
\begin{aligned}
p(x_n|Y_{n-1}) &= \int p(x_n|x_{n-1})\, p(x_{n-1}|Y_{n-1})\, dx_{n-1}\\
&= \int \frac{1}{\sqrt{2\pi Q_n}} e^{-\frac{(x_n - F_n\, x_{n-1})^2}{2Q_n}} \frac{1}{\sqrt{2\pi P_{n-1|n-1}}} e^{-\frac{(x_{n-1} - m_{n-1|n-1})^2}{2P_{n-1|n-1}}}\, dx_{n-1}, &(2.40)
\end{aligned}
$$

which is a convolution of Gaussian functions and results in a Gaussian predicted state distribution

$$p(x_n|Y_{n-1}) \equiv \mathcal{N}(x_n; m_{n|n-1}, P_{n|n-1}), \tag{2.41}$$

with mean and variance given by

$$
\begin{aligned}
m_{n|n-1} &= F_n\, m_{n-1|n-1}, & \text{(2.42a)}\\
P_{n|n-1} &= F_n^2\, P_{n-1|n-1} + Q_n. & \text{(2.42b)}
\end{aligned}
$$

The update step (2.33b) is now a product of Gaussian functions

$$
\begin{aligned}
p(x_n|Y_n) &\propto \ p(y_n|x_n)\, p(x_n|Y_{n-1})\\
&= \frac{1}{\sqrt{2\pi R_n}}\, e^{-\frac{(y_n - H_n x_n)^2}{2R_n}}\, \frac{1}{\sqrt{2\pi P_{n|n-1}}}\, e^{-\frac{(x_n - m_{n|n-1})^2}{2P_{n|n-1}}},
\end{aligned}
\tag{2.43}
$$

resulting in a Gaussian posterior pdf at time step n

$$p(x_n|Y_n) \equiv \mathcal{N}(x_n; m_{n|n}, P_{n|n}), \tag{2.44}$$

with mean and variance given by

$$
\begin{aligned}
m_{n|n} &= m_{n|n-1} + K_n\, (y_n - H_n\, m_{n|n-1}), & \text{(2.45a)}\\
P_{n|n} &= P_{n|n-1} - K_n\, H_n\, P_{n|n-1}, & \text{(2.45b)}
\end{aligned}
$$

where

$$K_n = \frac{P_{n|n-1}\, H_n}{H_n^2\, P_{n|n-1} + R_n}. \tag{2.46}$$

Since the initial state distribution $p(x_0)$ is Gaussian, under the stated conditions, all pdfs in the Bayesian recursion steps of Eq. (2.33) assume a Gaussian form. The Bayesian filtering iteration of Eq. (2.33) then simplifies to a recursive computation of the mean $m_{n|n}$ and variance $P_{n|n}$ of the Gaussian posterior pdf $p(x_n|Y_n)$ at each time step n:

$$\{m_{0|0}, P_{0|0}\} \xrightarrow{\text{predict}} \{m_{1|0}, P_{1|0}\} \xrightarrow{\text{update}} \{m_{1|1}, P_{1|1}\} \xrightarrow{\text{predict}} \ldots\ldots \xrightarrow{\text{update}} \{m_{n|n}, P_{n|n}\} \xrightarrow{\text{predict}} \ldots \tag{2.47}$$

via the set of equations

$$
\begin{aligned}
m_{n|n-1} &= F_n\, m_{n-1|n-1}, & \text{(2.48a)}\\
P_{n|n-1} &= F_n^2\, P_{n-1|n-1} + Q_n, & \text{(2.48b)}\\
S_n &= H_n^2\, P_{n|n-1} + R_n, & \text{(2.48c)}\\
K_n &= P_{n|n-1}\, H_n / S_n, & \text{(2.48d)}\\
m_{n|n} &= m_{n|n-1} + K_n\, (y_n - H_n\, m_{n|n-1}), & \text{(2.48e)}\\
P_{n|n} &= P_{n|n-1} - K_n\, H_n\, P_{n|n-1}, & \text{(2.48f)}
\end{aligned}
$$

for $n = 1, 2, \ldots$, which is the 1-D Kalman filter [6, 12].

In what follows, the Kalman filter will be described formally as the estimator of the parameters of a dynamical system whose evolution and measurement can be modeled using a linear Gaussian state space representation. Examples are provided, as well as extensions to nonlinear problems and distributed systems.

CHAPTER 3

The Kalman Filter

3.1 THEORY

The *Kalman filter* [3, 4] is the Bayesian solution to the problem of sequentially estimating the states of a dynamical system in which the state evolution and measurement processes are both linear and Gaussian. Consider a state space model of the form

$$\mathbf{x}_n = \mathbf{F}_n \mathbf{x}_{n-1} + \mathbf{v}_n, \tag{3.1a}$$
$$\mathbf{y}_n = \mathbf{H}_n \mathbf{x}_n + \mathbf{w}_n, \tag{3.1b}$$

where $\mathbf{F}_n$ is the $D \times D$ state-transition matrix, $\mathbf{v}_n$ is a $D \times 1$ Gaussian random state noise vector with zero mean and covariance matrix $\mathbf{Q}_n$, $\mathbf{H}_n$ is the $M \times D$ measurement matrix, and $\mathbf{w}_n$ is a $M \times 1$ Gaussian random measurement noise vector with zero mean and covariance matrix $\mathbf{R}_n$. The state space model (3.1) describes a dynamical system in which the state evolution and measurement processes are both linear and Gaussian. The conditional pdfs $p(\mathbf{x}_n | \mathbf{x}_{n-1})$ and $p(\mathbf{y}_n | \mathbf{x}_n)$ of the state and measurement vectors, respectively, are then given by the Gaussian pdfs

$$p(\mathbf{x}_n | \mathbf{x}_{n-1}) \equiv \mathcal{N}(\mathbf{x}_n; \mathbf{F}_n \mathbf{x}_{n-1}, \mathbf{Q}_n), \tag{3.2a}$$
$$p(\mathbf{y}_n | \mathbf{x}_n) \equiv \mathcal{N}(\mathbf{y}_n; \mathbf{H}_n \mathbf{x}_n, \mathbf{R}_n), \tag{3.2b}$$

where the notation $\mathcal{N}(\cdot\,; \mathbf{m}, \mathbf{P})$ is used to denote a Gaussian pdf with mean $\mathbf{m}$ and covariance $\mathbf{P}$. Suppose that the initial state distribution $p(\mathbf{x}_0)$ is also Gaussian. Following the sequential Bayesian estimation procedure described in Section 2.4, under the stated conditions, the posterior pdf $p(\mathbf{x}_n | \mathbf{Y}_n)$ at time step n can be shown to be Gaussian:

$$p(\mathbf{x}_n | \mathbf{Y}_n) \equiv \mathcal{N}(\mathbf{x}_n; \mathbf{m}_{n|n}, \mathbf{P}_{n|n}), \tag{3.3}$$

where $\mathbf{m}_{n|n}$ and $\mathbf{P}_{n|n}$ denote the Gaussian posterior mean and covariance at time step n. This is a direct consequence of all pdfs in the Bayesian recursion steps of Eq. (2.33) assuming a Gaussian form. The subscript notation '$n|n$' is used to indicate association with the (Gaussian) pdf of the state vector $\mathbf{x}_n$ at time step n computed using the measurements up to time step n. The Bayesian filtering iteration of Eq. (2.33) then reduces to a recursive computation of the mean $\mathbf{m}_{n|n}$ and covariance $\mathbf{P}_{n|n}$ of the Gaussian posterior pdf $p(\mathbf{x}_n | \mathbf{Y}_n)$ at each time step n:

$$\{\mathbf{m}_{0|0}, \mathbf{P}_{0|0}\} \xrightarrow{\text{predict}} \{\mathbf{m}_{1|0}, \mathbf{P}_{1|0}\} \xrightarrow{\text{update}} \{\mathbf{m}_{1|1}, \mathbf{P}_{1|1}\} \xrightarrow{\text{predict}} \ldots\ldots \xrightarrow{\text{update}} \{\mathbf{m}_{n|n}, \mathbf{P}_{n|n}\} \xrightarrow{\text{predict}} \ldots \tag{3.4}$$

The Kalman filter equations are given by [6, 12]

$$\begin{aligned}
\mathbf{m}_{n|n-1} &= \mathbf{F}_n\,\mathbf{m}_{n-1|n-1}, & \text{(3.5a)} \\
\mathbf{P}_{n|n-1} &= \mathbf{F}_n\,\mathbf{P}_{n-1|n-1}\,\mathbf{F}_n^T + \mathbf{Q}_n, & \text{(3.5b)} \\
\mathbf{S}_n &= \mathbf{H}_n\,\mathbf{P}_{n|n-1}\,\mathbf{H}_n^T + \mathbf{R}_n, & \text{(3.5c)} \\
\mathbf{K}_n &= \mathbf{P}_{n|n-1}\,\mathbf{H}_n^T\,\mathbf{S}_n^{-1}, & \text{(3.5d)} \\
\mathbf{m}_{n|n} &= \mathbf{m}_{n|n-1} + \mathbf{K}_n\,(\mathbf{y}_n - \mathbf{H}_n\,\mathbf{m}_{n|n-1}), & \text{(3.5e)} \\
\mathbf{P}_{n|n} &= \mathbf{P}_{n|n-1} - \mathbf{K}_n\,\mathbf{H}_n\,\mathbf{P}_{n|n-1}, & \text{(3.5f)}
\end{aligned}$$

for $n = 1, 2, \ldots$. Here, Eqs. (3.5a)–(3.5b) represent the prediction step and Eqs. (3.5c)–(3.5f) comprise the update step. Note that these equations are the multivariate versions of the 1-D Kalman filter equations (2.48) in Subsection 2.4.1. The Kalman filtering algorithm is summarized below.

Algorithm 1 Kalman filtering

Input: Linear Gaussian state space model (3.1), and a set of measurements $\{\mathbf{y}_1, \mathbf{y}_2, \ldots, \mathbf{y}_N\}$.
Output: Estimates of the states $\mathbf{x}_1, \mathbf{x}_2, \ldots, \mathbf{x}_N$ at time steps $n = 1, 2, \ldots, N$.

Initialize the mean and covariance of the Gaussian state distribution $\mathcal{N}(\mathbf{x}_0; \mathbf{m}_{0|0}, \mathbf{P}_{0|0})$ at time step $n = 0$.

For time steps $n = 1, 2, \ldots, N$, perform the following:

1. Carry out the prediction step

$$\begin{aligned}
\mathbf{m}_{n|n-1} &= \mathbf{F}_n\,\mathbf{m}_{n-1|n-1}, \\
\mathbf{P}_{n|n-1} &= \mathbf{F}_n\,\mathbf{P}_{n-1|n-1}\,\mathbf{F}_n^T + \mathbf{Q}_n.
\end{aligned}$$

2. Compute the estimate $\hat{\mathbf{x}}_n = \mathbf{m}_{n|n}$ of the state $\mathbf{x}_n$ using the update step

$$\begin{aligned}
\mathbf{S}_n &= \mathbf{H}_n\,\mathbf{P}_{n|n-1}\,\mathbf{H}_n^T + \mathbf{R}_n, \\
\mathbf{K}_n &= \mathbf{P}_{n|n-1}\,\mathbf{H}_n^T\,\mathbf{S}_n^{-1}, \\
\mathbf{m}_{n|n} &= \mathbf{m}_{n|n-1} + \mathbf{K}_n\,(\mathbf{y}_n - \mathbf{H}_n\,\mathbf{m}_{n|n-1}), \\
\mathbf{P}_{n|n} &= \mathbf{P}_{n|n-1} - \mathbf{K}_n\,\mathbf{H}_n\,\mathbf{P}_{n|n-1}.
\end{aligned}$$

3.2 IMPLEMENTATION

3.2.1 SAMPLE MATLAB CODE

We give a sample MATLAB implementation of the Kalman filter. The MATLAB function `kalman_filter.m` shown in Listing 3.1 performs one iteration of the Kalman filter prediction and update steps (Eq. (3.5)). This function will be used in all of the Kalman filtering examples

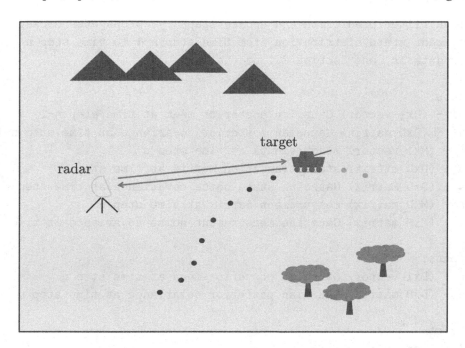

Figure 3.1: Generic scenario of a static ground-based radar system tracking a moving ground target. Information about the target's dynamics is combined with delay-Doppler (range and range-rate) measurements obtained from the radar to compute an estimate of the target's position and velocity.

given in Section 3.3. Note the use of the MATLAB '/' operator in implementing the update step (3.5d), that avoids explicit computation of the inverse of the matrix $\mathbf{S}_n$.

3.2.2 COMPUTATIONAL ISSUES

The matrix-vector operations in the Kalman filter prediction and update steps in general involve $O(D^2)$, $O(MD)$, and $O(M^3)$ complexity. The computational complexity, storage requirements, and numerical stability, can be optimized by exploiting structure. For example, the covariance matrices are symmetric and positive-semidefinite and the Cholesky factorization [17] can be utilized to represent and maintain them in the square root form $\mathbf{P} = \mathbf{L}\mathbf{L}^T$, where $\mathbf{L}$ is a lower triangular matrix [6, 12]. In addition, depending on the application, it may be possible to improve the

```
1   function [m_n,P_n] = kalman_filter (m_n1,P_n1,y_n,F_n,Q_n,H_n,R_n)
2
3   % Kalman filter prediction and update steps --- propagates Gaussian
4   % posterior state distribution from time step n-1 to time step n
5   % (for details, see Section 3.1 of the text)
6   %
7   % Inputs:
8   % m_n1 - (Dx1 vector) Gaussian posterior mean at time step n-1
9   % P_n1 - (DxD matrix) Gaussian posterior covariance at time step n-1
10  % y_n - (Mx1 vector) measurements at time step n
11  % F_n - (DxD matrix) state-transition matrix at time step n
12  % Q_n - (DxD matrix) Gaussian state noise covariance at time step n
13  % H_n - (MxD matrix) measurement matrix at time step n
14  % R_n - (MxM matrix) Gaussian measurement noise covariance at time step n
15  %
16  % Outputs:
17  % m_n - (Dx1 vector) Gaussian posterior mean at time step n
18  % P_n - (DxD matrix) Gaussian posterior covariance at time step n
19
20  % Predict
21  m_nn1 = F_n*m_n1;
22  P_nn1 = F_n*P_n1*F_n' + Q_n;
23
24  % Update
25  S_n = H_n*P_nn1*H_n' + R_n;
26  K_n = P_nn1*H_n'/S_n;
27  m_n = m_nn1 + K_n*(y_n-H_n*m_nn1);
28  P_n = P_nn1 - K_n*H_n*P_nn1;
```

Listing 3.1: kalman_filter.m

efficiency further by leveraging sparsity and the availability of any fast transforms (eg. FFT). It should be noted that the cost remains the same for each Kalman iteration.

3.3 EXAMPLES

The Kalman filter finds use in a wide variety of applications, e.g., target tracking [6], guidance and navigation [7–9], biomedical signal processing [18], motion estimation [19], and communications systems [20]. We now present several examples showing Kalman filtering in action.

3.3.1 TARGET TRACKING WITH RADAR

We consider the problem of radar-based tracking of a moving target. Figure 3.1 depicts a generic scenario in which a static ground-based radar system tracks a moving ground target.

In our example, we let the target move in the x-direction only, and denote with x_n the target's x-position at time step n and with $\dot{x}_n$ its x-velocity at time step n. Assuming that the target moves with approximately constant velocity, the linear dynamical model of the target's motion can be written as

$$
\begin{aligned}
x_n &= x_{n-1} + \Delta t\, \dot{x}_{n-1} + v_{n,1}, & \text{(3.6a)} \\
\dot{x}_n &= \dot{x}_{n-1} + v_{n,2}, & \text{(3.6b)}
\end{aligned}
$$

where Δt is the time interval between time steps $n-1$ and n, and $v_{n,1}$ and $v_{n,2}$ are Gaussian random perturbations with zero mean and covariances $E[v_{n,1}^2] = \sigma_x^2 \frac{(\Delta t)^3}{3}$, $E[v_{n,1} v_{n,2}] = \sigma_x^2 \frac{(\Delta t)^2}{2}$, and $E[v_{n,2}^2] = \sigma_x^2 \Delta t$ [6]. In matrix notation,

$$
\begin{bmatrix} x_n \\ \dot{x}_n \end{bmatrix} = \begin{bmatrix} 1 & \Delta t \\ 0 & 1 \end{bmatrix} \begin{bmatrix} x_{n-1} \\ \dot{x}_{n-1} \end{bmatrix} + \begin{bmatrix} v_{n,1} \\ v_{n,2} \end{bmatrix}. \tag{3.7}
$$

Thus, the state vector $\mathbf{x}_n = [x_n\ \dot{x}_n]^T$ is two-dimensional ($D = 2$), the state-transition matrix $\mathbf{F}_n$ is 2×2 and given by

$$
\mathbf{F}_n = \begin{bmatrix} 1 & \Delta t \\ 0 & 1 \end{bmatrix}, \tag{3.8}
$$

and the state noise vector $\mathbf{v}_n = [v_{n,1}\ v_{n,2}]^T$ is 2×1 and Gaussian with zero mean and covariance matrix

$$
\mathbf{Q}_n = \sigma_x^2 \begin{bmatrix} \frac{(\Delta t)^3}{3} & \frac{(\Delta t)^2}{2} \\ \frac{(\Delta t)^2}{2} & \Delta t \end{bmatrix}. \tag{3.9}
$$

The radar system utilizes time-delay and Doppler shift of transmitted and received signals to obtain noisy measurements of the target's range r_n (position x_n) and range-rate $\dot{r}_n$ (velocity $\dot{x}_n$). The measurement process can be written as

$$
\begin{aligned}
r_n &= x_n + w_{n,1}, & \text{(3.10a)} \\
\dot{r}_n &= \dot{x}_n + w_{n,2}, & \text{(3.10b)}
\end{aligned}
$$

where $w_{n,1}$ and $w_{n,2}$ are i.i.d. Gaussian with zero mean and variance σ_y^2. In matrix notation,

$$
\begin{bmatrix} r_n \\ \dot{r}_n \end{bmatrix} = \begin{bmatrix} x_n \\ \dot{x}_n \end{bmatrix} + \begin{bmatrix} w_{n,1} \\ w_{n,2} \end{bmatrix}.
\tag{3.11}
$$

Thus, the measurement vector $\mathbf{y}_n = [r_n \ \dot{r}_n]^T$ is two-dimensional ($M = 2$), the measurement matrix $\mathbf{H}_n = \mathbf{I}$ is the 2×2 identity matrix, and the measurement noise vector $\mathbf{w}_n = [w_{n,1} \ w_{n,2}]^T$ is 2×1 and Gaussian with zero mean and covariance matrix $\mathbf{R}_n = \sigma_y^2 \mathbf{I}$.

The state evolution and measurement models are both linear and Gaussian, and the Kalman filter can be used to estimate the moving target's position and velocity. The MATLAB program target_tracking.m shown in Listing 3.2 performs $N = 100$ time steps of target tracking using the Kalman filter. The state space model parameters were $\Delta t = 0.5, \sigma_x = 0.1, \sigma_y = 0.1$, the initial target position and velocity were set as $\mathbf{x}_0 = [1 \ 1]^T$, and the Kalman filter was initialized with Gaussian mean $\mathbf{m}_{0|0} = \mathbf{x}_0$ and covariance $\mathbf{P}_{0|0} = \mathbf{I}$. The MATLAB function kalman_filter.m shown in Listing 3.1 is utilized here to carry out the Kalman filter prediction and update steps (Eq. (3.5)). Figure 3.2 shows a plot of the actual and estimated target position and velocity. It can be seen that the Kalman filter is able to track the moving target accurately.

Target Tracking in Higher Dimensions

We now consider the radar-based tracking of a target moving in both x- and y- directions, and denote with (x_n, y_n) the target's position at time step n and with $(\dot{x}_n, \dot{y}_n)$ its velocity at time step n. Assuming that the target moves with approximately constant velocity, the linear dynamical model of the target's motion can be written in matrix notation as

$$
\begin{bmatrix} x_n \\ y_n \\ \dot{x}_n \\ \dot{y}_n \end{bmatrix} = \begin{bmatrix} 1 & 0 & \Delta t & 0 \\ 0 & 1 & 0 & \Delta t \\ 0 & 0 & 1 & 0 \\ 0 & 0 & 0 & 1 \end{bmatrix} \begin{bmatrix} x_{n-1} \\ y_{n-1} \\ \dot{x}_{n-1} \\ \dot{y}_{n-1} \end{bmatrix} + \begin{bmatrix} v_{n,1} \\ v_{n,2} \\ v_{n,3} \\ v_{n,4} \end{bmatrix},
\tag{3.12}
$$

where $v_{n,1}$ through $v_{n,4}$ are zero mean correlated Gaussian random perturbations as before. In this case, the state vector $\mathbf{x}_n = [x_n \ y_n \ \dot{x}_n \ \dot{y}_n]^T$ is four-dimensional ($D = 4$), the state-transition matrix $\mathbf{F}_n$ is 4×4 and given by

$$
\mathbf{F}_n = \begin{bmatrix} 1 & 0 & \Delta t & 0 \\ 0 & 1 & 0 & \Delta t \\ 0 & 0 & 1 & 0 \\ 0 & 0 & 0 & 1 \end{bmatrix},
\tag{3.13}
$$

```
1    % Radar based Target Tracking using the Kalman Filter
2
3    % State space model
4    delt = 0.5;                      % Time step interval
5    F_n = [1 delt; 0 1];             % State-transition matrix
6    sigma_x = 0.1; Q_n = sigma_x^2*[delt^3/3 delt^2/2;...
7        delt^2/2 delt];              % State noise covariance matrix
8    H_n = eye(2);                    % Measurement matrix
9    sigma_y = 0.1;                   % Measurement noise standard deviation
10   R_n = sigma_y^2*eye(2);          % Measurement noise covariance matrix
11
12   % Initialization
13   x(1:2,1) = [1; 1];               % Initial target x-position and x-velocity
14   m(1:2,1) = x(1:2,1);             % Gaussian posterior mean at time step 1
15   P(1:2,1:2,1) = eye(2);           % Gaussian posterior covariance at time step 1
16
17   % Track target using Kalman filter
18   for n = 2 : 100,                 % Time steps
19
20       % State propagation
21       v_n = mvnrnd([0 0],Q_n)';            % State noise vector
22       x(1:2,n) = F_n*x(1:2,n-1) + v_n;     % Markov linear Gaussian evolution
23
24       % Generate measurements
25       w_n = sigma_y*randn(2,1);            % Measurement noise vector
26       y_n = H_n*x(1:2,n) + w_n;            % Linear Gaussian measurements
27
28       % Compute Gaussian posterior mean and covariance at time step n
29       [m(1:2,n),P(1:2,1:2,n)] = kalman_filter (m(1:2,n-1),...
30           P(1:2,1:2,n-1),y_n,F_n,Q_n,H_n,R_n);
31   end
32
33   % Plot actual and estimated target position and velocity
34   figure, plot(x(1,:),x(2,:),'b'); hold on, plot(m(1,:),m(2,:),'r--');
35   xlabel('target x-position'); ylabel('target x-velocity');
36   title('Radar based target tracking using the Kalman filter');
37   legend('actual','estimated');
```

Listing 3.2: target_tracking.m

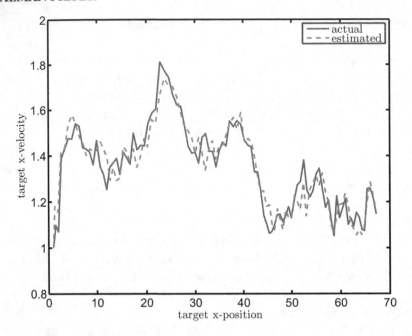

Figure 3.2: Radar-based target tracking using the Kalman filter. The target starts at x-position $x_0 = 1$ with x-velocity $\dot{x}_0 = 1$ and moves according the dynamical law given in (3.7), with $\Delta t = 0.5$ and $\sigma_x = 0.1$. Noisy range and range-rate measurements are collected as specified in (3.11), with $\sigma_y = 0.1$. $N = 100$ time steps of Kalman filtering are performed and the actual and estimated target position and velocity are shown in the plot.

and the state noise vector $\mathbf{v}_n = [v_{n,1} \ v_{n,2} \ v_{n,3} \ v_{n,4}]^T$ is 4×1 and Gaussian with zero mean and covariance matrix

$$\mathbf{Q}_n = \sigma_x^2 \begin{bmatrix} \frac{(\Delta t)^3}{3} & 0 & \frac{(\Delta t)^2}{2} & 0 \\ 0 & \frac{(\Delta t)^3}{3} & 0 & \frac{(\Delta t)^2}{2} \\ \frac{(\Delta t)^2}{2} & 0 & \Delta t & 0 \\ 0 & \frac{(\Delta t)^2}{2} & 0 & \Delta t \end{bmatrix}. \tag{3.14}$$

The measurement process collects information about the target's range r_n, range-rate $\dot{r}_n$, and bearing angle θ_n, and is now nonlinear and of the form

$$r_n = \sqrt{x_n^2 + y_n^2} + w_{n,1}, \tag{3.15a}$$

$$\dot{r}_n = \frac{x_n \dot{x}_n + y_n \dot{y}_n}{\sqrt{x_n^2 + y_n^2}} + w_{n,2}, \tag{3.15b}$$

$$\theta_n = \tan^{-1}\left(\frac{y_n}{x_n}\right) + w_{n,3}, \tag{3.15c}$$

where the radar position is taken as the origin $(0, 0)$, and $w_{n,1}$, $w_{n,2}$, and $w_{n,3}$ are i.i.d. Gaussian with zero mean and variance σ_y^2. In matrix notation,

$$\begin{bmatrix} r_n \\ \dot{r}_n \\ \theta_n \end{bmatrix} = \mathbf{h} \left(\begin{bmatrix} x_n \\ y_n \\ \dot{x}_n \\ \dot{y}_n \end{bmatrix} \right) + \begin{bmatrix} w_{n,1} \\ w_{n,2} \\ w_{n,3} \end{bmatrix}. \tag{3.16}$$

Thus, the measurement vector $\mathbf{y}_n = [r_n \ \dot{r}_n \ \theta_n]^T$ is three-dimensional ($M = 3$), the function $\mathbf{h} : \mathbb{R}^4 \mapsto \mathbb{R}^3$ describes the nonlinear measurement-state relationship, and the measurement noise vector $\mathbf{w}_n = [w_{n,1} \ w_{n,2} \ w_{n,3}]^T$ is 3×1 and Gaussian with zero mean and covariance matrix $\mathbf{R}_n = \sigma_y^2 \mathbf{I}$.

Due to the nonlinearity of the measurement model, the Kalman filter cannot be applied to estimate the moving target's position and velocity. However, techniques such as the extended Kalman filter [6, 12] discussed in Section 3.1 can be used to solve this estimation problem.

3.3.2   CHANNEL ESTIMATION IN COMMUNICATIONS SYSTEMS

In wireless communications systems operating in an urban environment, signals from the transmitter may not reach the receiver directly due to scattering. Typically, the received signal is a superposition of the direct (line-of-sight) arrival and several reflected, phase-shifted, and delayed signals. This effect is known as multipath propagation. Figure 3.3 depicts a multipath wireless communications channel in an urban environment. Multipath channels are commonly modeled as a tapped-delay line (finite impulse response or FIR digital filter). In orthogonal frequency-division multiplexing (OFDM) [21] systems, the time-varying properties of wireless communications channels have to be estimated. In the context of the Kalman filter, this problem can be posed as a system identification problem of estimating the FIR coefficients of the channel, measured by transmitting and receiving a discrete-time test signal filtered by the channel. A state space model describing the system can be obtained as follows.

Let the time-varying filter coefficients of the channel evolve according to an autoregressive process, given by

$$a_{n,d} = \alpha_d \, a_{n-1,d} + v_{n,d}, \quad d = 1, 2, 3, \tag{3.17}$$

where $a_{n,1}$, $a_{n,2}$, $a_{n,3}$ are the channel coefficients at time step n (the FIR filter is of order 3), α_1, α_2, α_3 are the autoregressive parameters, and $v_{n,1}$, $v_{n,2}$, $v_{n,3}$ are i.i.d. Gaussian with zero mean and variance σ_x^2. In matrix notation,

$$\begin{bmatrix} a_{n,1} \\ a_{n,2} \\ a_{n,3} \end{bmatrix} = \begin{bmatrix} \alpha_1 & 0 & 0 \\ 0 & \alpha_2 & 0 \\ 0 & 0 & \alpha_3 \end{bmatrix} \begin{bmatrix} a_{n-1,1} \\ a_{n-1,2} \\ a_{n-1,3} \end{bmatrix} + \begin{bmatrix} v_{n,1} \\ v_{n,2} \\ v_{n,3} \end{bmatrix}. \tag{3.18}$$

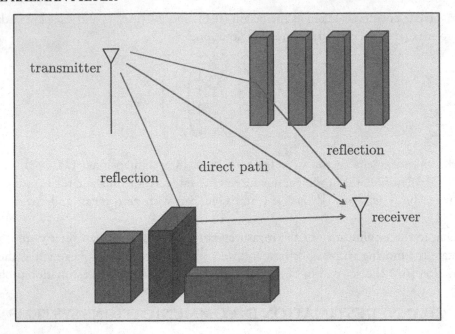

Figure 3.3: A multipath wireless communications channel in an urban environment. The received signal is a superposition of the direct arrival and several reflected, phase-shifted, and delayed signals. The multipath channel is modeled as an FIR filter, whose coefficients can be estimated by transmitting and receiving a test signal through the channel.

Thus, the state vector $\mathbf{x}_n = [a_{n,1}\ a_{n,2}\ a_{n,3}]^T$ is three-dimensional ($D = 3$), the state-transition matrix $\mathbf{F}_n$ is 3×3 and given by

$$\mathbf{F}_n = \begin{bmatrix} \alpha_1 & 0 & 0 \\ 0 & \alpha_2 & 0 \\ 0 & 0 & \alpha_3 \end{bmatrix}, \tag{3.19}$$

and the state noise vector $\mathbf{v}_n = [v_{n,1}\ v_{n,2}\ v_{n,3}]^T$ is 3×1 and Gaussian with zero mean and co-variance matrix $\mathbf{Q}_n = \sigma_x^2 \mathbf{I}$.

The noisy measurements are obtained as the filtered output for a test signal u_n propagated through the channel as

$$y_n = \sum_{d=1}^{3} a_{n,d}\, u_{n-d+1} + w_n, \tag{3.20}$$

where w_n is Gaussian with zero mean and variance σ_y^2. In matrix notation,

$$y_n = \begin{bmatrix} u_n & u_{n-1} & u_{n-2} \end{bmatrix} \begin{bmatrix} a_{n,1} \\ a_{n,2} \\ a_{n,3} \end{bmatrix} + w_n. \qquad (3.21)$$

Thus, the measurement vector $\mathbf{y}_n = y_n$ is one-dimensional ($M = 1$), the measurement matrix $\mathbf{H}_n = \mathbf{u}_n^T = [u_n \ u_{n-1} \ u_{n-2}]$ is 1×3, and the measurement noise vector $\mathbf{w}_n = w_n$ is scalar and Gaussian with zero mean and variance $\mathbf{R}_n = \sigma_y^2$.

The state evolution and measurement models are both linear and Gaussian, and the Kalman filter can be used to estimate the time-varying channel coefficients. The MATLAB program `channel_estimation.m` shown in Listing 3.3 performs $N = 500$ time steps of channel estimation using the Kalman filter. The state space model parameters were $\boldsymbol{\alpha} = [0.85 \ \ 1.001 \ \ -0.95]^T$, $\sigma_x = 0.1$, $u_n \overset{\text{i.i.d.}}{\sim} \mathcal{N}(u_n; 0, 1)$, $\sigma_y = 0.1$, the initial channel coefficients were $\mathbf{x}_0 \sim \mathcal{N}(\mathbf{x}_0; \mathbf{0}, \mathbf{I})$, and the Kalman filter was initialized with Gaussian mean $\mathbf{m}_{0|0} = \mathbf{x}_0$ and covariance $\mathbf{P}_{0|0} = \mathbf{I}$. The MATLAB function `kalman_filter.m` shown in Listing 3.1 is utilized to carry out the Kalman filter prediction and update steps (Eq. (3.5)). Figure 3.4c shows plots of the actual and estimated channel coefficients. It can be seen that the Kalman filter is able to estimate the time-varying channel coefficients with good accuracy.

```
1   % Channel Estimation in Communication Systems using the Kalman Filter
2
3   % State space model
4   alpha = [0.85; 1.001; -0.95];    % Autoregressive parameters
5   F_n = diag(alpha);               % State-transition matrix
6   sigma_x = 0.1;                   % State noise standard deviation
7   Q_n = sigma_x^2*eye(3);          % State noise covariance matrix
8   u = randn(501,1);                % Test signal
9   sigma_y = 0.1;                   % Measurement noise standard deviation
10  R_n = sigma_y^2;                 % Measurement noise covariance matrix
11
12  % Initialization
13  x(1:3,1) = randn(3,1);           % Initial channel coefficients
14  m(1:3,1) = x(1:3,1);             % Gaussian posterior mean at time step 1
15  P(1:3,1:3,1) = eye(3);           % Gaussian posterior covariance at time step 1
16
17  % Channel Estimation using the Kalman filter
18  for n = 2 : 500,                           % Time steps
19
20      % State propagation
21      v_n = sigma_x*randn(3,1);              % State noise vector
```

```
22      x(1:3,n) = F_n*x(1:3,n-1) + v_n;      % Markov linear Gaussian evolution
23
24      % Generate measurements
25      H_n = u(n+1:-1:n-1,1)';               % Measurement matrix
26      w_n = sigma_y*randn(1,1);             % Measurement noise vector
27      y_n = H_n*x(1:3,n) + w_n;             % Linear Gaussian measurements
28
29      % Compute Gaussian posterior mean and covariance at time step n
30      [m(1:3,n),P(1:3,1:3,n)] = kalman_filter (m(1:3,n-1),...
31          P(1:3,1:3,n-1),y_n,F_n,Q_n,H_n,R_n);
32  end
33
34  % Plot actual and estimated channel coefficients
35  n = [0:499]'; figure, plot(n,x(1,:),'b'); hold on, plot(n,m(1,:),'r--');
36  xlabel('time'); ylabel('coefficient 1'); legend('actual','estimated');
37  title('Channel Estimation in Communication Systems using Kalman Filter');
38  figure, plot(n,x(2,:),'g'); hold on, plot(n,m(2,:),'m--');
39  xlabel('time'); ylabel('coefficient 2'); legend('actual','estimated');
40  title('Channel Estimation in Communication Systems using Kalman Filter');
41  figure, plot(n,x(3,:),'k'); hold on, plot(n,m(3,:),'c--');
42  xlabel('time'); ylabel('coefficient 3'); legend('actual','estimated');
43  title('Channel Estimation in Communication Systems using Kalman Filter');
```

Listing 3.3: `channel_estimation.m`

3.3.3   RECURSIVE LEAST SQUARES (RLS) ADAPTIVE FILTERING

We consider the adaptive filtering problem where the goal is to design a digital filter that operates linearly on an input signal such that the filtered output closely approximates a desired signal of interest. The filter constantly adapts in time based on the statistics of the given signals, which can be non-stationary (in the stationary case, the optimum filter is time-invariant and given by the Wiener filter as shown in Section 2.2). Adaptive filters are used in many applications, including signal denoising, prediction, and system modeling and identification [12]. The adaptive filtering problem can be formulated in the state space setting as follows.

Assume that the time-varying filter coefficients evolve according to a dynamical law, given by

$$a_{n,d} = \lambda^{-1/2} a_{n-1,d}, \quad d = 1, \dots, D, \tag{3.22}$$

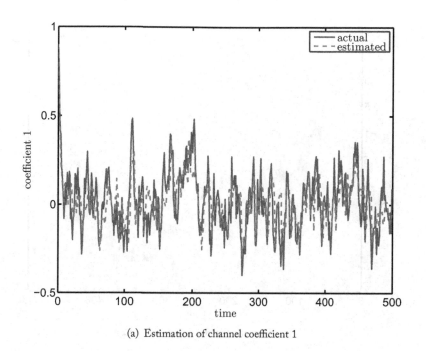

(a) Estimation of channel coefficient 1

Figure 3.4a: Channel estimation in communication systems using the Kalman filter. The $D = 3$ channel coefficients evolve according the dynamical law given in (3.18), with $\alpha = [0.85\ \ 1.001\ \ -0.95]^T$ and $\sigma_x = 0.1$. Noisy measurements are collected as specified in (3.21), with $u_n \overset{\text{i.i.d.}}{\sim} \mathcal{N}(u_n; 0, 1)$ and $\sigma_y = 0.1$. $N = 500$ time steps of Kalman filtering are performed and the actual and estimated channel coefficients are shown in the plots.

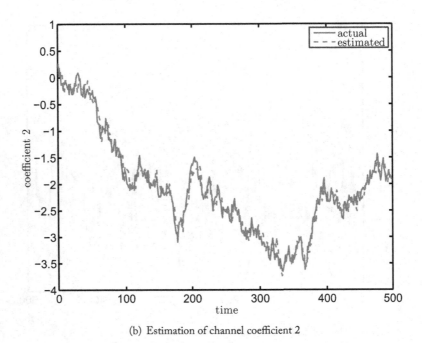

(b) Estimation of channel coefficient 2

Figure 3.4b: Channel estimation in communication systems using the Kalman filter. The $D = 3$ channel coefficients evolve according the dynamical law given in (3.18), with $\alpha = [0.85 \ 1.001 \ -0.95]^T$ and $\sigma_x = 0.1$. Noisy measurements are collected as specified in (3.21), with $u_n \overset{\text{i.i.d.}}{\sim} \mathcal{N}(u_n; 0, 1)$ and $\sigma_y = 0.1$. $N = 500$ time steps of Kalman filtering are performed and the actual and estimated channel coefficients are shown in the plots.

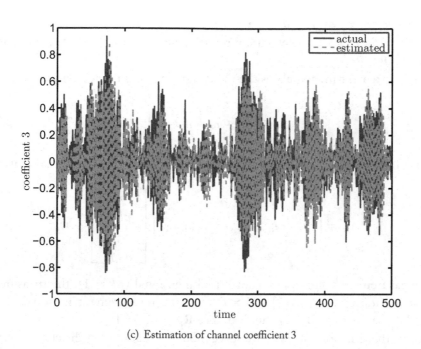

(c) Estimation of channel coefficient 3

Figure 3.4c: Channel estimation in communication systems using the Kalman filter. The $D = 3$ channel coefficients evolve according the dynamical law given in (3.18), with $\alpha = [0.85 \ 1.001 \ -0.95]^T$ and $\sigma_x = 0.1$. Noisy measurements are collected as specified in (3.21), with $u_n \overset{\text{i.i.d.}}{\sim} \mathcal{N}(u_n; 0, 1)$ and $\sigma_y = 0.1$. $N = 500$ time steps of Kalman filtering are performed and the actual and estimated channel coefficients are shown in the plots.

where $a_{n,1}, a_{n,2}, \ldots, a_{n,D}$ are the filter coefficients at time step n (the filter is FIR and of order D) and $\lambda \in (0, 1]$ is an exponential weight parameter. In matrix notation,

$$\begin{bmatrix} a_{n,1} \\ a_{n,2} \\ \vdots \\ a_{n,D} \end{bmatrix} = \lambda^{-1/2} \, \mathbf{I} \begin{bmatrix} a_{n-1,1} \\ a_{n-1,2} \\ \vdots \\ a_{n-1,D} \end{bmatrix}. \tag{3.23}$$

Thus, the state vector $\mathbf{x}_n = [a_{n,1} \; a_{n,2} \; \cdots \; a_{n,D}]^T$ is D-dimensional, and the state-transition matrix $\mathbf{F}_n = \lambda^{-1/2} \mathbf{I}$ is $D \times D$. The evolution model is deterministic and the state noise covariance matrix is $\mathbf{Q}_n = \mathbf{0}$.

The measurement model relates the input signal u_n and filtered output y_n as

$$y_n = \sum_{d=1}^{D} a_{n,d} \, u_{n-d+1} + w_n, \tag{3.24}$$

where w_n is Gaussian with zero mean and unit variance ($\sigma_y^2 = 1$). In matrix notation,

$$y_n = \begin{bmatrix} u_n & u_{n-1} & \cdots & u_{n-D+1} \end{bmatrix} \begin{bmatrix} a_{n,1} \\ a_{n,2} \\ \vdots \\ a_{n,D} \end{bmatrix} + w_n. \tag{3.25}$$

Thus, the measurement vector $\mathbf{y}_n = y_n$ is one-dimensional ($M = 1$), the measurement matrix $\mathbf{H}_n = \mathbf{u}_n^T = [u_n \; u_{n-1} \; \cdots \; u_{n-D+1}]$ is $1 \times D$, and the measurement noise vector $\mathbf{w}_n = w_n$ is scalar and Gaussian with zero mean and variance $\mathbf{R}_n = \sigma_y^2 = 1$.

Observe the similarity of this state space model to that used in the channel estimation system identification example of Subsection 3.3.2. For this model, the Kalman filter equations (3.5) can be simplified to

$$\mathbf{k}_n = \frac{\mathbf{P}_{n-1|n-1} \, \mathbf{u}_n}{\mathbf{u}_n^T \, \mathbf{P}_{n-1|n-1} \, \mathbf{u}_n + \lambda}, \tag{3.26a}$$

$$\mathbf{m}_{n|n} = \lambda^{-1/2} \mathbf{m}_{n-1|n-1} + \mathbf{k}_n \left(y_n - \lambda^{-1/2} \mathbf{u}_n^T \mathbf{m}_{n-1|n-1} \right), \tag{3.26b}$$

$$\mathbf{P}_{n|n} = \lambda^{-1} \mathbf{P}_{n-1|n-1} - \lambda^{-1} \mathbf{k}_n \mathbf{u}_n^T \mathbf{P}_{n-1|n-1}, \tag{3.26c}$$

for $n = 1, 2, \ldots$. In particular, the substitution $\mathbf{m}_{n|n} = \lambda^{-n/2} \hat{\mathbf{a}}_n$ (adaptive filter) and $y_n = \lambda^{-n/2} s_n$ (desired signal) yields

$$\mathbf{k}_n = \frac{\mathbf{P}_{n-1|n-1} \, \mathbf{u}_n}{\mathbf{u}_n^T \, \mathbf{P}_{n-1|n-1} \, \mathbf{u}_n + \lambda}, \tag{3.27a}$$

$$\hat{\mathbf{a}}_n = \hat{\mathbf{a}}_{n-1} + \mathbf{k}_n \left(s_n - \mathbf{u}_n^T \hat{\mathbf{a}}_{n-1} \right), \tag{3.27b}$$

$$\mathbf{P}_{n|n} = \lambda^{-1} \mathbf{P}_{n-1|n-1} - \lambda^{-1} \mathbf{k}_n \mathbf{u}_n^T \mathbf{P}_{n-1|n-1}, \tag{3.27c}$$

for $n = 1, 2, \ldots$, which are also known as the *recursive least squares (RLS) filtering* equations [12, 22–26]. Figure 3.5 shows a block diagram of the adaptive filtering framework.

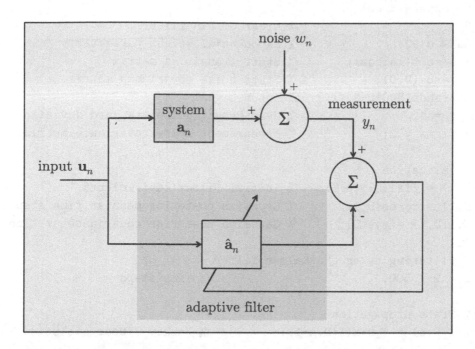

Figure 3.5: Block diagram of the adaptive filtering framework. The filter operates linearly on an input signal and constantly adapts such that the filtered output closely matches the given measurements.

The state evolution and measurement models are linear and Gaussian, and the Kalman filter (or, equivalently, the RLS filter) can be used to estimate the time-varying filter coefficients. The MATLAB program `rls_filtering.m` shown in Listing 3.4 performs $N = 500$ time steps of estimation using the Kalman filter. The state space model parameters were $D = 32$, $\lambda = 0.99$, $u_n \overset{\text{i.i.d.}}{\sim} \mathcal{N}(u_n; 0, 100)$, $\sigma_y = 1$ (these values were selected similar to the example in MATLAB's `adaptfilt.rls` function documentation [27]), and the Kalman filter was initialized with Gaussian mean $\mathbf{m}_{0|0} = \mathbf{0}$ and covariance $\mathbf{P}_{0|0} = \mathbf{I}$. The MATLAB function `kalman_filter.m` shown in Listing 3.1 is utilized to carry out the Kalman filter prediction and update steps (Eq. (3.5)). Figure 3.6 shows a plot of the actual and estimated filter coefficients. It can be seen that the Kalman filter is able to estimate the time-varying filter coefficients with good accuracy.

The Kalman filter described in this chapter can be applied for parameter estimation of dynamical systems that may be represented using a linear Gaussian state space model. Several extensions to the standard setup are possible, including estimation algorithms for nonlinear systems, systems with non-Gaussian noise, and distributed systems involving data collected from multi-

```matlab
% Recursive least squares (RLS) adaptive filtering using the Kalman filter

% State space model
D = 32;                          % Order of FIR filter
lambda = 0.99;                   % Exponential weight parameter
F_n = 1/sqrt(lambda);           % State-transition matrix
Q_n = zeros(D,D);               % State noise covariance matrix
u = 10*randn(500+D-2,1);        % Input signal
sigma_y = 1;                     % Measurement noise standard deviation
R_n = sigma_y^2;                 % Measurement noise covariance matrix

% Initialization
x(1:D,1) = fir1(D-1,0.5);        % Initial filter coefficients
m(1:D,1) = zeros(D,1);           % Gaussian posterior mean at time step 1
P(1:D,1:D,1) = eye(D);           % Gaussian posterior covariance at time step 1

% RLS filtering using the Kalman filter
for n = 2 : 500,                 % Time steps

    % State propagation
    x(1:D,n) = F_n*x(1:D,n-1);          % Markov linear evolution

    % Generate measurements
    H_n = u(n+D-2:-1:n-1,1)';            % Measurement matrix
    w_n = sigma_y*randn(1,1);            % Measurement noise vector
    y_n = H_n*x(1:D,n) + w_n;            % Linear Gaussian measurements

    % Compute Gaussian posterior mean and covariance at time step n
    [m(1:D,n),P(1:D,1:D,n)] = kalman_filter (m(1:D,n-1),...
        P(1:D,1:D,n-1),y_n,F_n,Q_n,H_n,R_n);
end

% Plot actual and estimated filter coefficients
figure, stem(x(1:D,500),'b'); hold on, stem(m(1:D,500),'r--');
xlabel('coefficient number'); ylabel('filter coefficient');
title('RLS adaptive filtering using the Kalman filter');
legend('actual','estimated');
```

Listing 3.4: rls_filtering.m

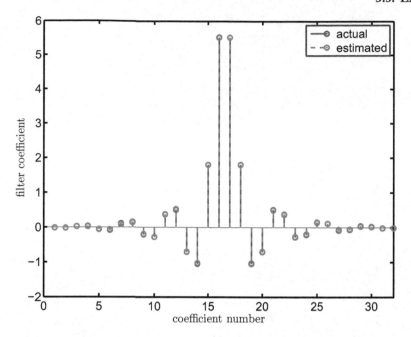

Figure 3.6: RLS adaptive filtering using the Kalman filter. The $D = 32$ order FIR filter coefficients evolve according the dynamical law given in (3.23), with $\lambda = 0.99$. Noisy measurements are collected as specified in (3.25), with $u_n \overset{\text{i.i.d.}}{\sim} \mathcal{N}(u_n; 0, 100)$ and $\sigma_y = 1$. $N = 500$ time steps of Kalman filtering are performed and the actual and estimated filter coefficients at the final time step are shown in the plot.

ple sources. In the following chapter, we will discuss the extended Kalman filter for nonlinear problems, as well as a decentralized formulation of the Kalman filter.

Figure 5.6.

CHAPTER 4

Extended and Decentralized Kalman Filtering

We complete our discussion of Kalman filtering with an outline of two popular extensions developed for nonlinear problems and distributed applications. Nonlinear state space models are often encountered, for example, in target tracking [6] and inertial navigation systems [8, 9]. Distributed estimation tasks arise frequently in the context of sensor networks [28] and multisensor tracking [29, 30]. In this chapter we examine the extended and decentralized Kalman filters and illustrate their utility through examples.

4.1 EXTENDED KALMAN FILTER

When the state evolution and measurement processes are nonlinear and Gaussian, the *extended Kalman filter* [6, 12] can be utilized for estimation of the states. The state space model for this case can be written as

$$
\begin{aligned}
\mathbf{x}_n &= \mathbf{f}(\mathbf{x}_{n-1}) + \mathbf{v}_n, & (4.1a) \\
\mathbf{y}_n &= \mathbf{h}(\mathbf{x}_n) + \mathbf{w}_n, & (4.1b)
\end{aligned}
$$

where $\mathbf{f} : \mathbb{R}^D \mapsto \mathbb{R}^D$ is the state-transition function, $\mathbf{h} : \mathbb{R}^D \mapsto \mathbb{R}^M$ is the measurement function, and $\mathbf{v}_n$ and $\mathbf{w}_n$ are independent Gaussian noise vectors as defined earlier. Assuming that $\mathbf{f}$ and $\mathbf{h}$ are differentiable, the model can be linearized by making use of the Jacobian matrices:

$$
\tilde{\mathbf{F}}_n = \left.\frac{\partial \mathbf{f}}{\partial \mathbf{x}}\right|_{\mathbf{m}_{n-1|n-1}} = \left.\begin{bmatrix} \frac{\partial f_1}{\partial x_1} & \frac{\partial f_1}{\partial x_2} & \cdots & \frac{\partial f_1}{\partial x_D} \\ \frac{\partial f_2}{\partial x_1} & \frac{\partial f_2}{\partial x_2} & \cdots & \frac{\partial f_2}{\partial x_D} \\ \vdots & \vdots & \ddots & \vdots \\ \frac{\partial f_D}{\partial x_1} & \frac{\partial f_D}{\partial x_2} & \cdots & \frac{\partial f_D}{\partial x_D} \end{bmatrix}\right|_{\mathbf{m}_{n-1|n-1}}, \qquad (4.2a)
$$

$$
\tilde{\mathbf{H}}_n = \left.\frac{\partial \mathbf{h}}{\partial \mathbf{x}}\right|_{\mathbf{m}_{n|n-1}} = \left.\begin{bmatrix} \frac{\partial h_1}{\partial x_1} & \frac{\partial h_1}{\partial x_2} & \cdots & \frac{\partial h_1}{\partial x_D} \\ \frac{\partial h_2}{\partial x_1} & \frac{\partial h_2}{\partial x_2} & \cdots & \frac{\partial h_2}{\partial x_D} \\ \vdots & \vdots & \ddots & \vdots \\ \frac{\partial h_M}{\partial x_1} & \frac{\partial h_M}{\partial x_2} & \cdots & \frac{\partial h_M}{\partial x_D} \end{bmatrix}\right|_{\mathbf{m}_{n|n-1}}. \qquad (4.2b)
$$

The extended Kalman filter equations are given by [6, 12]

$$
\begin{align}
\mathbf{m}_{n|n-1} &= \mathbf{f}(\mathbf{m}_{n-1|n-1}), \tag{4.3a} \\
\mathbf{P}_{n|n-1} &= \tilde{\mathbf{F}}_n \, \mathbf{P}_{n-1|n-1} \, \tilde{\mathbf{F}}_n^T + \mathbf{Q}_n, \tag{4.3b} \\
\mathbf{S}_n &= \tilde{\mathbf{H}}_n \, \mathbf{P}_{n|n-1} \, \tilde{\mathbf{H}}_n^T + \mathbf{R}_n, \tag{4.3c} \\
\mathbf{K}_n &= \mathbf{P}_{n|n-1} \, \tilde{\mathbf{H}}_n^T \, \mathbf{S}_n^{-1}, \tag{4.3d} \\
\mathbf{m}_{n|n} &= \mathbf{m}_{n|n-1} + \mathbf{K}_n \, (\mathbf{y}_n - \mathbf{h}(\mathbf{m}_{n|n-1})), \tag{4.3e} \\
\mathbf{P}_{n|n} &= \mathbf{P}_{n|n-1} - \mathbf{K}_n \, \tilde{\mathbf{H}}_n \, \mathbf{P}_{n|n-1}, \tag{4.3f}
\end{align}
$$

for $n = 1, 2, \ldots$, with Eqs. (4.3a)–(4.3b) representing the prediction step and Eqs. (4.3c)–(4.3f) comprising the update step. Note that the extended Kalman filter equations (4.3) reduce to the Kalman filter equations (3.5) when the state-transition function $\mathbf{f}$ and measurement function $\mathbf{h}$ are linear. The extended Kalman filtering algorithm is summarized below.

Algorithm 2 Extended Kalman filtering

Input: Nonlinear Gaussian state space model (4.1), and a set of measurements $\{\mathbf{y}_1, \mathbf{y}_2, \ldots, \mathbf{y}_N\}$.
Output: Estimates of the states $\mathbf{x}_1, \mathbf{x}_2, \ldots, \mathbf{x}_N$ at time steps $n = 1, 2, \ldots, N$.

Initialize the mean and covariance of the Gaussian state distribution $\mathcal{N}(\mathbf{x}_0; \mathbf{m}_{0|0}, \mathbf{P}_{0|0})$ at time step $n = 0$.

For time steps $n = 1, 2, \ldots, N$, perform the following:

1. Carry out the prediction step

$$
\begin{align}
\mathbf{m}_{n|n-1} &= \mathbf{f}(\mathbf{m}_{n-1|n-1}), \\
\mathbf{P}_{n|n-1} &= \tilde{\mathbf{F}}_n \, \mathbf{P}_{n-1|n-1} \, \tilde{\mathbf{F}}_n^T + \mathbf{Q}_n.
\end{align}
$$

2. Compute the estimate $\hat{\mathbf{x}}_n = \mathbf{m}_{n|n}$ of the state $\mathbf{x}_n$ using the update step

$$
\begin{align}
\mathbf{S}_n &= \tilde{\mathbf{H}}_n \, \mathbf{P}_{n|n-1} \, \tilde{\mathbf{H}}_n^T + \mathbf{R}_n, \\
\mathbf{K}_n &= \mathbf{P}_{n|n-1} \, \tilde{\mathbf{H}}_n^T \, \mathbf{S}_n^{-1}, \\
\mathbf{m}_{n|n} &= \mathbf{m}_{n|n-1} + \mathbf{K}_n \, (\mathbf{y}_n - \mathbf{h}(\mathbf{m}_{n|n-1})), \\
\mathbf{P}_{n|n} &= \mathbf{P}_{n|n-1} - \mathbf{K}_n \, \tilde{\mathbf{H}}_n \, \mathbf{P}_{n|n-1}.
\end{align}
$$

The MATLAB function `extended_kalman_filter.m` shown in Listing 4.1 performs one iteration of the extended Kalman filter prediction and update steps.

Other algorithms for nonlinear and/or non-Gaussian filtering include those based on unscented transforms [12, 31] and Monte Carlo techniques such as the particle filter [5, 16].

```
1   function [m_n,P_n] = extended_kalman_filter (m_n1,P_n1,y_n,f,Ft_n,Q_n...
2       ,h,Ht_n,R_n)
3
4   % Extended Kalman filter prediction and update steps --- propagates
5   % Gaussian posterior state distribution from time step n-1 to time step n
6   % (for details, see Section 4.1 of the text)
7   %
8   % Inputs:
9   % m_n1 - (Dx1 vector) Gaussian posterior mean at time step n-1
10  % P_n1 - (DxD matrix) Gaussian posterior covariance at time step n-1
11  % y_n - (Mx1 vector) measurements at time step n
12  % f(x) - (Dx1 vector function of Dx1 vector x) state-transition function
13  % Ft_n - (DxD matrix) state-transition Jacobian matrix at time step n
14  % Q_n - (DxD matrix) Gaussian state noise covariance at time step n
15  % h(x) - (Mx1 vector function of Dx1 vector x) measurement function
16  % Ht_n - (MxD matrix) measurement Jacobian matrix at time step n
17  % R_n - (MxM matrix) Gaussian measurement noise covariance at time step n
18  %
19  % Outputs:
20  % m_n - (Dx1 vector) Gaussian posterior mean at time step n
21  % P_n - (DxD matrix) Gaussian posterior covariance at time step n
22
23  % Predict
24  m_nn1 = feval(f,m_n1);
25  P_nn1 = Ft_n*P_n1*Ft_n' + Q_n;
26
27  % Update
28  S_n = Ht_n*P_nn1*Ht_n' + R_n;
29  K_n = P_nn1*Ht_n'/S_n;
30  m_n = m_nn1 + K_n*(y_n-feval(h,m_nn1));
31  P_n = P_nn1 - K_n*Ht_n*P_nn1;
```

Listing 4.1: extended_kalman_filter.m

4.1.1   EXAMPLE: PREDATOR-PREY SYSTEM

We consider a bio-system occupied by a predator population and a prey population. This type of a system is modeled using the Lotka-Volterra equations [32]. In this model, the predator and prey interact based on the following equations:

$$x_{n,1} = \left[1 + \Delta t \left(1 - \frac{x_{n-1,2}}{c_2}\right)\right] x_{n-1,1} + v_{n,1}, \tag{4.4a}$$

$$x_{n,2} = \left[1 - \Delta t \left(1 - \frac{x_{n-1,1}}{c_1}\right)\right] x_{n-1,2} + v_{n,2}. \tag{4.4b}$$

Here, $x_{n,1}$ and $x_{n,2}$ denote respectively the prey and predator populations at time step n, Δt is the time interval between time steps $n-1$ and n, c_1 and c_2 are constant model parameters, and $v_{n,1}$ and $v_{n,2}$ are i.i.d. Gaussian with zero mean and variance σ_x^2. Clearly, the two populations do not live in isolation. For example, if left with an infinite food supply and no predators, a population of rabbits would experience normal population growth. Similarly, a population of foxes with no rabbit population would face decline. When together, these populations interact according to (4.4). Figure 4.1 depicts the cycle of predator-prey population dynamics.

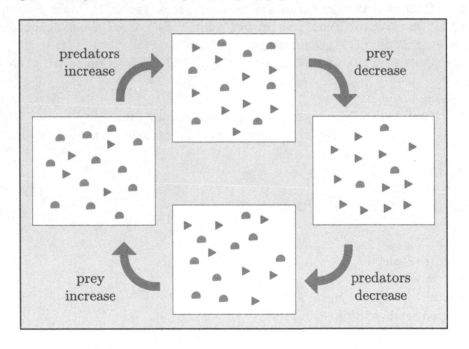

Figure 4.1: Cycle of predator-prey population dynamics. When there is an abundance of preys, the predator population increases. But this increase in predators drives the prey population down. With a scarcity of preys, the predator population is forced to decrease. The decrease in predators then results in prey population increasing, and the cycle continues.

In matrix notation, (4.4) can be expressed as

$$\begin{bmatrix} x_{n,1} \\ x_{n,2} \end{bmatrix} = \mathbf{f} \left(\begin{bmatrix} x_{n-1,1} \\ x_{n-1,2} \end{bmatrix} \right) + \begin{bmatrix} v_{n,1} \\ v_{n,2} \end{bmatrix}. \tag{4.5}$$

The state vector $\mathbf{x}_n = [x_{n,1}\ x_{n,2}]^T$ is two-dimensional ($D = 2$), $\mathbf{f} : \mathbb{R}^2 \mapsto \mathbb{R}^2$ is the nonlinear state-transition function, and the state noise vector $\mathbf{v}_n = [v_{n,1}\ v_{n,2}]^T$ is 2×1 and Gaussian with zero mean and covariance matrix $\mathbf{Q}_n = \sigma_x^2 \mathbf{I}$.

We wish to estimate the predator and prey populations based on noisy measurements of the total population

$$y_n = x_{n,1} + x_{n,2} + w_n, \tag{4.6}$$

where w_n is Gaussian with zero mean and variance σ_y^2. In matrix notation,

$$y_n = \begin{bmatrix} 1 & 1 \end{bmatrix} \begin{bmatrix} x_{n,1} \\ x_{n,2} \end{bmatrix} + w_n. \tag{4.7}$$

Thus, the measurement vector $\mathbf{y}_n = y_n$ is one-dimensional ($M = 1$), the measurement matrix $\mathbf{H}_n = [1\ 1]$ is 1×2, and the measurement noise vector $\mathbf{w}_n = w_n$ is scalar and Gaussian with zero mean and variance $\mathbf{R}_n = \sigma_y^2$.

The state space model above is nonlinear and Gaussian, and the extended Kalman filter can be used for estimation. In particular, the state-transition function Jacobian matrix (Eq. (4.2a)) is 2×2 and given by

$$\tilde{\mathbf{F}}_n = \begin{bmatrix} 1 + \Delta t \left(1 - \frac{m_{n-1|n-1,2}}{c_2} \right) & -\Delta t \, \frac{m_{n-1|n-1,1}}{c_2} \\ \Delta t \, \frac{m_{n-1|n-1,2}}{c_1} & 1 - \Delta t \left(1 - \frac{m_{n-1|n-1,1}}{c_1} \right) \end{bmatrix}, \tag{4.8}$$

where $\mathbf{m}_{n-1|n-1} = \begin{bmatrix} m_{n-1|n-1,1}\ m_{n-1|n-1,2} \end{bmatrix}^T$ is the mean of the Gaussian posterior pdf at time step $n - 1$.

The MATLAB program predator_prey.m shown in Listing 4.2 performs $N = 2000$ time steps of predator-prey population estimation using the extended Kalman filter. The state space model parameters were $\Delta t = 0.01$, $c_1 = 300$, $c_2 = 200$, $\sigma_x = 1$, $\sigma_y = 10$, the initial prey and predator populations were set as $\mathbf{x}_0 = [400\ 100]^T$ (these values were selected to generate a plot similar to that in [33, Ch. 16]), and the extended Kalman filter was initialized with Gaussian mean $\mathbf{m}_{0|0} = \mathbf{x}_0$ and covariance $\mathbf{P}_{0|0} = 100\,\mathbf{I}$. The MATLAB function extended_kalman_filter.m shown in Listing 4.1 is utilized to carry out the extended Kalman filter prediction and update steps (Eq. (4.3)). Figure 4.2b shows a plot of the actual and estimated predator-prey populations. It can be seen that the extended Kalman filter is able to track the predator-prey dynamics with good accuracy.

```matlab
% Solution of Predator-Prey Equations using the Extended Kalman Filter

function predator_prey

% State space model
delt = 0.01;                    % Time step interval
c1 = 300; c2 = 200;             % Predator-Prey model parameters
sigma_x = 1;                    % State noise standard deviation
Q_n = sigma_x^2*eye(2);         % State noise covariance matrix
H_n = [1 1];                    % Measurement matrix
sigma_y = 10;                   % Measurement noise standard deviation
R_n = sigma_y^2;                % Measurement noise covariance matrix

% Initialization
x(1:2,1) = [400; 100];          % Initial prey and predator populations
m(1:2,1) = x(1:2,1);            % Gaussian posterior mean at time step 1
P(1:2,1:2,1) = 100*eye(2);      % Gaussian posterior covariance at time step 1

% Predator-Prey estimation using extended Kalman filter
for n = 2 : 2000,                       % Time steps

    % State propagation
    v_n = sigma_x*randn(2,1);           % State noise vector
    x(1:2,n) = f(x(1:2,n-1)) + v_n;     % Markov nonlinear Gaussian evolution

    % Generate measurements
    w_n = sigma_y*randn(1,1);           % Measurement noise vector
    y_n = H_n*x(1:2,n) + w_n;           % Linear Gaussian measurements

    % State-transition function Jacobian matrix
    Ft_n = [1+delt*(1-m(2,n-1)/c2) -delt*m(1,n-1)/c2;...
        delt*m(2,n-1)/c1 1-delt*(1-m(1,n-1)/c1)];

    % Compute Gaussian posterior mean and covariance at time step n
    [m(1:2,n),P(1:2,1:2,n)] = extended_kalman_filter (m(1:2,n-1),...
        P(1:2,1:2,n-1),y_n,@f,Ft_n,Q_n,@h,H_n,R_n);
end
```

```matlab
39   % Plot actual and estimated predator-prey populations
40   % Evolution in time
41   t = [0:1999]'*delt;
42   figure, plot(t,x(1,:),'b'); hold on, plot(t,m(1,:),'r--');
43   hold on, plot(t,x(2,:),'g.-'); hold on, plot(t,m(2,:),'m-.');
44   xlabel('time'); ylabel('population');
45   title('Solution of predator-prey equations using extended Kalman filter');
46   legend('actual preys','estimated preys','actual predators',...
47       'estimated predators');
48   % Phase plot
49   figure, plot(x(1,:),x(2,:),'b'); hold on, plot(m(1,:),m(2,:),'r--');
50   xlabel('prey population'); ylabel('predator population');
51   title('Solution of predator-prey equations using extended Kalman filter');
52   legend('actual','estimated');
53
54   % State-transition function
55   function x_n = f (x_n1)
56       x_n = [(1+delt*(1-x_n1(2,1)/c2))*x_n1(1,1); ...
57           (1-delt*(1-x_n1(1,1)/c1))*x_n1(2,1)];
58   end
59
60   % Measurement function
61   function y_n = h (x_n)
62       y_n = x_n(1,1) + x_n(2,1);
63   end
64
65   end
```

Listing 4.2: `predator_prey.m`

4.2   DECENTRALIZED KALMAN FILTERING

In distributed multisensor network configurations, the *decentralized Kalman filter* [34] enables estimation for linear Gaussian state space models without the need for a central communication and processing facility. The resulting estimation framework is robust to failure of one or more nodes and can yield performance benefits such as faster processing capability. In this scheme, each sensing node first collects measurements about the system's state and communicates a small amount of information with the other nodes. Next, each node locally processes the information received from the other nodes in conjunction with its own data to compute an overall estimate

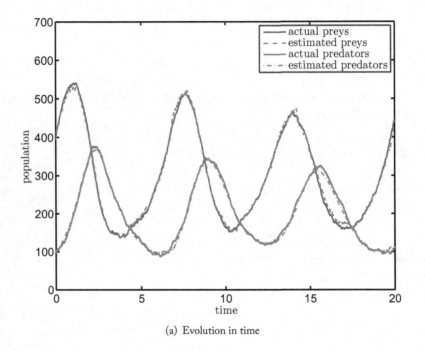

(a) Evolution in time

Figure 4.2a: Solution of the Predator-Prey equations using the extended Kalman filter. The initial prey and predator populations are 400 and 100, respectively, and evolve according the dynamical law given in (4.4), with $c_1 = 300$ and $c_2 = 200$. Noisy total population measurements are collected as specified in (4.6), with $\sigma_y = 10$. $N = 2000$ time steps of extended Kalman filtering are performed and the actual and estimated predator-prey populations are shown in the plots.

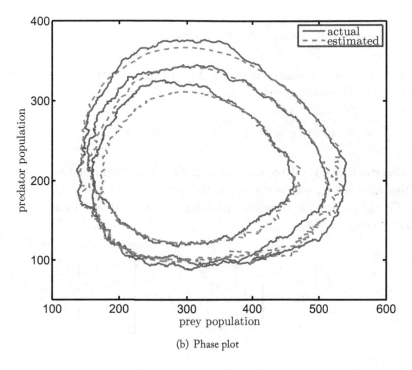

(b) Phase plot

Figure 4.2b: Solution of the Predator-Prey equations using the extended Kalman filter. The initial prey and predator populations are 400 and 100, respectively, and evolve according the dynamical law given in (4.4), with $c_1 = 300$ and $c_2 = 200$. Noisy total population measurements are collected as specified in (4.6), with $\sigma_y = 10$. $N = 2000$ time steps of extended Kalman filtering are performed and the actual and estimated predator-prey populations are shown in the plots.

of the state. In the following brief description, it is assumed for simplicity that measurement validation is not used and that the communication between the nodes is synchronized; the reader is referred to [34] for the general method.

Using matrix inversion lemmas and some algebra, the Kalman filter update equations (3.5e)–(3.5f) for the mean $\mathbf{m}_{n|n}$ and covariance $\mathbf{P}_{n|n}$ of the (Gaussian) posterior pdf $p(\mathbf{x}_n|Y_n)$ at time step n can be rewritten in *information form* as

$$\mathbf{P}_{n|n}^{-1} = \mathbf{P}_{n|n-1}^{-1} + \mathbf{H}_n^T \mathbf{R}_n^{-1} \mathbf{H}_n, \tag{4.9a}$$

$$\mathbf{P}_{n|n}^{-1} \mathbf{m}_{n|n} = \mathbf{P}_{n|n-1}^{-1} \mathbf{m}_{n|n-1} + \mathbf{H}_n^T \mathbf{R}_n^{-1} \mathbf{y}_n. \tag{4.9b}$$

Let K be the number of distributed nodes used to collect measurements, with the measurement equation for each of the nodes expressed as

$$\mathbf{y}_{n,k} = \mathbf{H}_{n,k} \mathbf{x}_n + \mathbf{w}_{n,k}, \quad k = 1, \ldots, K, \tag{4.10}$$

where $\mathbf{y}_{n,k}$ denotes the $M_k \times 1$ vector of measurements collected by sensor node k at time step n, $\mathbf{H}_{n,k}$ is the $M_k \times D$ measurement matrix for node k, $\mathbf{x}_n$ is the $D \times 1$ state vector, and $\mathbf{w}_{n,k}$ is a $M_k \times 1$ Gaussian random measurement noise vector with zero mean and covariance matrix $\mathbf{R}_{n,k}$. In block matrix notation,

$$\begin{bmatrix} \mathbf{y}_{n,1} \\ \mathbf{y}_{n,2} \\ \vdots \\ \mathbf{y}_{n,K} \end{bmatrix} = \begin{bmatrix} \mathbf{H}_{n,1} \\ \mathbf{H}_{n,2} \\ \vdots \\ \mathbf{H}_{n,K} \end{bmatrix} \mathbf{x}_n + \begin{bmatrix} \mathbf{w}_{n,1} \\ \mathbf{w}_{n,2} \\ \vdots \\ \mathbf{w}_{n,K} \end{bmatrix}, \tag{4.11}$$

or

$$\mathbf{y}_n = \mathbf{H}_n \mathbf{x}_n + \mathbf{w}_n, \tag{4.12}$$

where $\mathbf{y}_n = [\mathbf{y}_{n,1}^T \, \mathbf{y}_{n,2}^T \, \cdots \, \mathbf{y}_{n,K}^T]^T$ is the combined $M \times 1$ measurement vector ($M = \sum_{k=1}^K M_k$), $\mathbf{H}_n = [\mathbf{H}_{n,1}^T \, \mathbf{H}_{n,2}^T \, \cdots \, \mathbf{H}_{n,K}^T]^T$ is the combined $M \times D$ measurement matrix, and $\mathbf{w}_n = [\mathbf{w}_{n,1}^T \, \mathbf{w}_{n,2}^T \, \cdots \, \mathbf{w}_{n,K}^T]^T$ is the combined $M \times 1$ Gaussian random measurement noise vector with zero mean and covariance matrix

$$\mathbf{R}_n = \begin{bmatrix} \mathbf{R}_{n,1} & 0 & \cdots & 0 \\ 0 & \mathbf{R}_{n,2} & \cdots & 0 \\ \vdots & \vdots & \ddots & \vdots \\ 0 & 0 & \cdots & \mathbf{R}_{n,K} \end{bmatrix}. \tag{4.13}$$

Here, the measurement noise vectors of the different sensor nodes have been assumed to be uncorrelated. Due to the block structure of the measurement model, we can write

$$\mathbf{H}_n^T \mathbf{R}_n^{-1} \mathbf{H}_n \;=\; \sum_{k=1}^{K} \mathbf{H}_{n,k}^T \mathbf{R}_{n,k}^{-1} \mathbf{H}_{n,k}, \tag{4.14a}$$

$$\mathbf{H}_n^T \mathbf{R}_n^{-1} \mathbf{y}_n \;=\; \sum_{k=1}^{K} \mathbf{H}_{n,k}^T \mathbf{R}_{n,k}^{-1} \mathbf{y}_{n,k}. \tag{4.14b}$$

Substituting into (4.9) yields

$$\mathbf{P}_{n|n}^{-1} \;=\; \mathbf{P}_{n|n-1}^{-1} + \sum_{k=1}^{K} \mathbf{H}_{n,k}^T \mathbf{R}_{n,k}^{-1} \mathbf{H}_{n,k}, \tag{4.15a}$$

$$\mathbf{P}_{n|n}^{-1} \mathbf{m}_{n|n} \;=\; \mathbf{P}_{n|n-1}^{-1} \mathbf{m}_{n|n-1} + \sum_{k=1}^{K} \mathbf{H}_{n,k}^T \mathbf{R}_{n,k}^{-1} \mathbf{y}_{n,k}, \tag{4.15b}$$

which represent the decentralized Kalman filter update equations [34]. The quantities inside the summation terms in (4.15) need to be communicated between the nodes in order to perform the state update step.

Note that the decentralized Kalman filter update step (4.15) is mathematically equivalent to the (centralized) Kalman filter update step in Eqs. (3.5c)–(3.5f); no approximation has been used. Assuming that the same global state evolution model of Eq. (3.1a) (and therefore the same prediction step in Eqs. (3.5a)–(3.5b)) is used by all nodes, with identical initialization, the decentralized Kalman filter provides exactly the same state estimates as the centralized Kalman filter.

Computationally, the decentralized Kalman filter can be viewed as a parallelization of the centralized Kalman filter, in which the central measurement process has been divided between the K distributed sensor nodes. Since (a) the local measurement model at each node is smaller (dimension M_k for node k) than the global measurement model (dimension M), (b) the overhead associated with the combination stage (summations in (4.15)) is small, and (c) the nodes perform their computations in parallel, significant savings are afforded in terms of processing speed (ignoring communication delays).

The decentralized Kalman filtering algorithm is summarized below.

4.2.1 EXAMPLE: DISTRIBUTED OBJECT TRACKING

We consider the tracking of a moving object based on information collected using multiple sensor nodes. Such a task can arise during satellite and cellular network-based navigation, as depicted in Figure 4.3. We are interested in estimating the object's position and velocity in a distributed manner, without making use of a central processor.

Let (x_n, y_n) denote the object's position at time step n and $(\dot{x}_n, \dot{y}_n)$ its velocity at time step n, so that the state vector $\mathbf{x}_n = [x_n \; y_n \; \dot{x}_n \; \dot{y}_n]^T$ is four-dimensional ($D = 4$). Assume that

---

Algorithm 3 Decentralized Kalman filtering

Input: Linear Gaussian state evolution model (3.1a), linear Gaussian measurement model (4.10), and a set of measurements $\{\mathbf{y}_{1,k}, \mathbf{y}_{2,k}, \ldots, \mathbf{y}_{N,k}\}$ at each of K distributed nodes ($k = 1, 2, \ldots, K$).
Output: Estimates of the states $\mathbf{x}_1, \mathbf{x}_2, \ldots, \mathbf{x}_N$ at time steps $n = 1, 2, \ldots, N$, at each node k.

Initialize the mean and covariance of the Gaussian state distribution $\mathcal{N}(\mathbf{x}_0; \mathbf{m}_{0|0}, \mathbf{P}_{0|0})$ at time step $n = 0$, at each node k.

For time steps $n = 1, 2, \ldots, N$, at each node k perform the following:

1. Carry out the prediction step

$$
\begin{aligned}
\mathbf{m}_{n|n-1} &= \mathbf{F}_n\, \mathbf{m}_{n-1|n-1}, \\
\mathbf{P}_{n|n-1} &= \mathbf{F}_n\, \mathbf{P}_{n-1|n-1}\, \mathbf{F}_n^T + \mathbf{Q}_n.
\end{aligned}
$$

2. Calculate the quantities $\mathbf{H}_{n,k}^T\, \mathbf{R}_{n,k}^{-1}\, \mathbf{H}_{n,k}$ and $\mathbf{H}_{n,k}^T\, \mathbf{R}_{n,k}^{-1}\, \mathbf{y}_{n,k}$ and communicate to all other nodes.

3. Compute the estimate $\hat{\mathbf{x}}_n = \mathbf{m}_{n|n}$ of the state $\mathbf{x}_n$ using the update step

$$
\mathbf{P}_{n|n} = \left(\mathbf{P}_{n|n-1}^{-1} + \sum_{k=1}^{K} \mathbf{H}_{n,k}^T\, \mathbf{R}_{n,k}^{-1}\, \mathbf{H}_{n,k} \right)^{-1},
$$

$$
\mathbf{m}_{n|n} = \mathbf{P}_{n|n} \left(\mathbf{P}_{n|n-1}^{-1}\, \mathbf{m}_{n|n-1} + \sum_{k=1}^{K} \mathbf{H}_{n,k}^T\, \mathbf{R}_{n,k}^{-1}\, \mathbf{y}_{n,k} \right).
$$

---

the object moves according to the approximately constant velocity model given in (3.12) (see Subsection 3.3.1).

In this example, the measurement system is comprised of $K = 3$ sensor nodes, with node 1 measuring the object's x-position, node 2 measuring the object's y-position, and node 3 measuring the object's x-position and x-velocity. Thus, the dimensionality of the local measurements is $M_1 = 1$, $M_2 = 1$, and $M_3 = 2$, and the measurement vectors are

$$
\begin{aligned}
\mathbf{y}_{n,1} &= \mathbf{H}_{n,1}\, \mathbf{x}_n + \mathbf{w}_{n,1}, & \text{(4.16a)} \\
\mathbf{y}_{n,2} &= \mathbf{H}_{n,2}\, \mathbf{x}_n + \mathbf{w}_{n,2}, & \text{(4.16b)} \\
\mathbf{y}_{n,3} &= \mathbf{H}_{n,3}\, \mathbf{x}_n + \mathbf{w}_{n,3}, & \text{(4.16c)}
\end{aligned}
$$

with the measurement matrices $\mathbf{H}_{n,1} = [1\ 0\ 0\ 0]$, $\mathbf{H}_{n,2} = [0\ 1\ 0\ 0]$, and $\mathbf{H}_{n,3} = \left[\begin{smallmatrix} 1 & 0 & 0 & 0 \\ 0 & 0 & 1 & 0 \end{smallmatrix}\right]$, and the measurement noise covariance matrices $\mathbf{R}_{n,1} = \sigma_y^2$, $\mathbf{R}_{n,2} = \sigma_y^2$, and $\mathbf{R}_{n,3} = \sigma_y^2\, \mathbf{I}$. Note that the object's y-velocity is not measured by any of the sensing nodes.

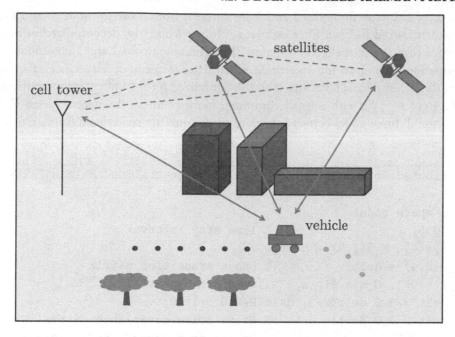

Figure 4.3: A satellite and cellular network-based navigation scenario. The automobile's position and velocity are estimated based on radio signals from navigation satellites within range and cellular network data where available.

The decentralized Kalman filter is now used to estimate the moving object's position and velocity. The MATLAB program `distributed_tracking.m` shown in Listing 4.3 performs $N = 100$ time steps of distributed tracking using the decentralized Kalman filter. The state space model parameters were $\Delta t = 0.5$, $\sigma_x = 0.1$, $\sigma_y = 0.25$, and the initial object position and velocity were set as $\mathbf{x}_0 = [1\ 1\ 0.1\ 0.1]^T$. Decentralized Kalman filtering is performed at each of the three nodes, and is initialized with Gaussian mean $\mathbf{m}_{0|0} = \mathbf{0}$ and covariance $\mathbf{P}_{0|0} = \mathbf{I}$. For time steps $n < 40$ and $n \geq 70$, perfect communication occurs between all nodes. For time steps $40 \leq n < 70$, nodes 1 and 2 have no communication with node 3. Figure 4.4c shows plots of the actual and estimated object position and the mean square error (MSE) in position and velocity. The MSE is computed using Monte Carlo simulation of 1 million trajectories, with object states initialized as $\mathbf{x}_0 \sim \mathcal{N}(\mathbf{x}_0; \mathbf{0}, \mathbf{I})$. For comparison, the performance of the centralized Kalman filter (using the same data) implemented in the MATLAB function `kalman_filter.m` shown in Listing 3.1 is also given.

It can be seen that decentralized Kalman filtering is able to track the moving object accurately. For time steps $n < 40$, the decentralized Kalman filter estimates for all three nodes coincide with those from the centralized Kalman filter. This is expected, as the filters have the same information, initialization, prediction, and update. For time steps $40 \leq n < 70$, the decentralized

Kalman filter estimates for nodes 1 and 2 are different from those for node 3 (and also different from the centralized Kalman filter estimates). In particular, the decentralized estimates have a higher error than the centralized estimates. This is because nodes 1 and 2 communicate their local information to compute the decentralized estimates, while node 3 uses its local information to compute the decentralized estimates (and the centralized Kalman filter still uses all information). For time steps $n \geq 70$, with normal communications restored, the decentralized Kalman filter estimates for all three nodes converge back to those from the centralized Kalman filter.

```
1   % Distributed Object Tracking using the Decentralized Kalman Filter
2
3   % State space model
4   delt = 0.5;                         % Time step interval
5   F_n = eye(4); F_n(1,3) = delt;...
6       F_n(2,4) = delt;               % State-transition matrix
7   sigma_x = 0.1; Q_n = sigma_x^2*[delt^3/3 0 delt^2/2 0;...
8       0 delt^3/3 0 delt^2/2; delt^2/2 0 delt 0;...
9       0 delt^2/2 0 delt];            % State noise covariance matrix
10  sigma_y = 0.25;                     % Measurement noise standard deviation
11  H_n1 = [1 0 0 0];                   % Node 1 measurement matrix
12  R_n1 = sigma_y^2;                   % Node 1 measurement noise covariance matrix
13  H_n2 = [0 1 0 0];                   % Node 2 measurement matrix
14  R_n2 = sigma_y^2;                   % Node 2 measurement noise covariance matrix
15  H_n3 = [1 0 0 0; 0 0 1 0];          % Node 3 measurement matrix
16  R_n3 = sigma_y^2*eye(2);            % Node 3 measurement noise covariance matrix
17  H_n = [H_n1; H_n2; H_n3];           % Central measurement matrix
18  R_n = blkdiag(R_n1,R_n2,R_n3);      % Central measurement noise covariance matrix
19
20  % Initialization
21  x(1:4,1) = [1; 1; 0.1; 0.1];        % Initial object position and velocity
22  m1(1:4,1) = zeros(4,1);             % Initial node 1 Gaussian posterior mean
23  P1(1:4,1:4,1) = eye(4);             % Initial node 1 Gaussian posterior covariance
24  m2(1:4,1) = zeros(4,1);             % Initial node 2 Gaussian posterior mean
25  P2(1:4,1:4,1) = eye(4);             % Initial node 2 Gaussian posterior covariance
26  m3(1:4,1) = zeros(4,1);             % Initial node 3 Gaussian posterior mean
27  P3(1:4,1:4,1) = eye(4);             % Initial node 3 Gaussian posterior covariance
28  m(1:4,1) = zeros(4,1);              % Initial central Gaussian posterior mean
29  P(1:4,1:4,1) = eye(4);              % Initial central Gaussian posterior covariance
30
31  % Track object using the decentralized Kalman filter
```

```
32   for n = 2 : 100,                    % Time steps
33
34       % State propagation
35       v_n = mvnrnd([0 0 0 0],Q_n)';   % State noise vector
36       x(1:4,n) = F_n*x(1:4,n-1) + v_n; % Markov linear Gaussian evolution
37
38       % Generate measurements
39       w_n1 = sigma_y*randn(1,1);      % Node 1 measurement noise vector
40       y_n1 = H_n1*x(1:4,n) + w_n1;    % Node 1 linear Gaussian measurements
41       w_n2 = sigma_y*randn(1,1);      % Node 2 measurement noise vector
42       y_n2 = H_n2*x(1:4,n) + w_n2;    % Node 2 linear Gaussian measurements
43       w_n3 = sigma_y*randn(2,1);      % Node 3 measurement noise vector
44       y_n3 = H_n3*x(1:4,n) + w_n3;    % Node 3 linear Gaussian measurements
45       y_n = [y_n1; y_n2; y_n3];       % Central linear Gaussian measurements
46
47       if (n<=40 || n>70)
48
49           % Perfect communication between all nodes
50
51           % Decentralized Kalman filter computations at node 1
52           % Predict
53           m1_nn1 = F_n*m1(1:4,n-1);
54           P1_nn1 = F_n*P1(1:4,1:4,n-1)*F_n' + Q_n;
55           % Update
56           P1(1:4,1:4,n) = inv(inv(P1_nn1) + H_n1'*(R_n1\H_n1) +...
57               H_n2'*(R_n2\H_n2) + H_n3'*(R_n3\H_n3));
58           m1(1:4,n) = P1(1:4,1:4,n)*(P1_nn1\m1_nn1 + H_n1'*...
59               (R_n1\y_n1) + H_n2'*(R_n2\y_n2) + H_n3'*(R_n3\y_n3));
60
61           % Decentralized Kalman filter computations at node 2
62           % Predict
63           m2_nn1 = F_n*m2(1:4,n-1);
64           P2_nn1 = F_n*P2(1:4,1:4,n-1)*F_n' + Q_n;
65           % Update
66           P2(1:4,1:4,n) = inv(inv(P2_nn1) + H_n1'*(R_n1\H_n1) +...
67               H_n2'*(R_n2\H_n2) + H_n3'*(R_n3\H_n3));
68           m2(1:4,n) = P2(1:4,1:4,n)*(P2_nn1\m2_nn1 + H_n1'*...
69               (R_n1\y_n1) + H_n2'*(R_n2\y_n2) + H_n3'*(R_n3\y_n3));
70
```

```matlab
% Decentralized Kalman filter computations at node 3
% Predict
m3_nn1 = F_n*m3(1:4,n-1);
P3_nn1 = F_n*P3(1:4,1:4,n-1)*F_n' + Q_n;
% Update
P3(1:4,1:4,n) = inv(inv(P3_nn1) + H_n1'*(R_n1\H_n1) +...
    H_n2'*(R_n2\H_n2) + H_n3'*(R_n3\H_n3));
m3(1:4,n) = P3(1:4,1:4,n)*(P3_nn1\m3_nn1 + H_n1'*...
    (R_n1\y_n1) + H_n2'*(R_n2\y_n2) + H_n3'*(R_n3\y_n3));

else

    % Nodes 1 and 2 have no communication with node 3

    % Decentralized Kalman filter computations at node 1
    % Predict
    m1_nn1 = F_n*m1(1:4,n-1);
    P1_nn1 = F_n*P1(1:4,1:4,n-1)*F_n' + Q_n;
    % Update
    P1(1:4,1:4,n) = inv(inv(P1_nn1) + H_n1'*(R_n1\H_n1) +...
        H_n2'*(R_n2\H_n2));
    m1(1:4,n) = P1(1:4,1:4,n)*(P1_nn1\m1_nn1 + H_n1'*...
        (R_n1\y_n1) + H_n2'*(R_n2\y_n2));

    % Decentralized Kalman filter computations at node 2
    % Predict
    m2_nn1 = F_n*m2(1:4,n-1);
    P2_nn1 = F_n*P2(1:4,1:4,n-1)*F_n' + Q_n;
    % Update
    P2(1:4,1:4,n) = inv(inv(P2_nn1) + H_n1'*(R_n1\H_n1) +...
        H_n2'*(R_n2\H_n2));
    m2(1:4,n) = P2(1:4,1:4,n)*(P2_nn1\m2_nn1 + H_n1'*...
        (R_n1\y_n1) + H_n2'*(R_n2\y_n2));

    % Decentralized Kalman filter computations at node 3
    % Predict
    m3_nn1 = F_n*m3(1:4,n-1);
    P3_nn1 = F_n*P3(1:4,1:4,n-1)*F_n' + Q_n;
    % Update
```

```
110        P3(1:4,1:4,n) = inv(inv(P3_nn1) + H_n3'*(R_n3\H_n3));
111        m3(1:4,n) = P3(1:4,1:4,n)*(P3_nn1\m3_nn1 + H_n3'*(R_n3\y_n3));
112
113     end
114
115     % Centralized Kalman filter computations
116     [m(1:4,n),P(1:4,1:4,n)] = kalman_filter (m(1:4,n-1),...
117        P(1:4,1:4,n-1),y_n,F_n,Q_n,H_n,R_n);
118 end
119
120 % Plot actual and estimated object position
121 figure, plot(x(1,:),x(2,:)); hold on, plot(m1(1,:),m1(2,:),'r');
122 plot(m2(1,:),m2(2,:),'g--'); plot(m3(1,:),m3(2,:),'m-.');
123 plot(m(1,:),m(2,:),'k:');
124 xlabel('object x-position'); ylabel('object y-position');
125 title('Distributed object tracking using the decentralized Kalman filter');
126 legend('actual','decentralized estimate (node 1)',...
127     'decentralized estimate (node 2)','decentralized estimate (node 3)',...
128     'centralized estimate');
```

Listing 4.3: distributed_tracking.m

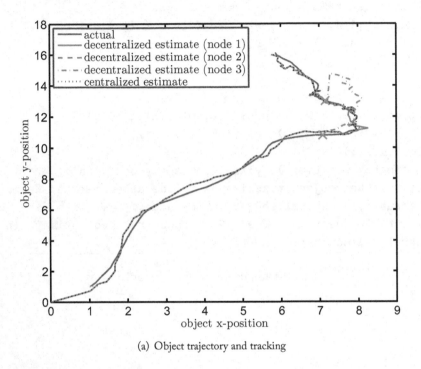

(a) Object trajectory and tracking

Figure 4.4a: Distributed object tracking using the decentralized Kalman filter. The object moves according the dynamical law given in (3.12), with $\Delta t = 0.5$ and $\sigma_x = 0.1$. Noisy measurements are collected using $K = 3$ sensor nodes as specified in (4.16), with $\sigma_y = 0.25$. $N = 100$ time steps of decentralized Kalman filtering are performed and the actual and estimated object position and the mean square error in position and velocity are shown in the plots. For time steps $n < 40$ and $n \geq 70$, perfect communication occurs between all nodes, and for time steps $40 \leq n < 70$, nodes 1 and 2 have no communication with node 3 (these time instants are indicated on the object's trajectory with cyan 'x's). For comparison, the performance of a centralized Kalman filter (using the same data) is also given.

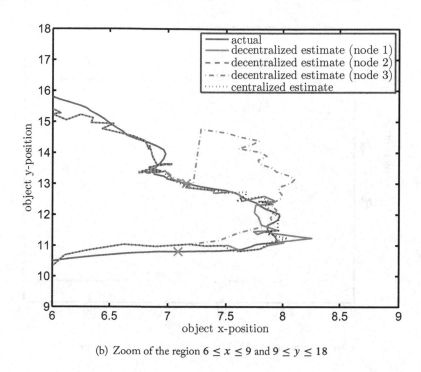

(b) Zoom of the region $6 \leq x \leq 9$ and $9 \leq y \leq 18$

Figure 4.4b: Distributed object tracking using the decentralized Kalman filter. The object moves according the dynamical law given in (3.12), with $\Delta t = 0.5$ and $\sigma_x = 0.1$. Noisy measurements are collected using $K = 3$ sensor nodes as specified in (4.16), with $\sigma_y = 0.25$. $N = 100$ time steps of decentralized Kalman filtering are performed and the actual and estimated object position and the mean square error in position and velocity are shown in the plots. For time steps $n < 40$ and $n \geq 70$, perfect communication occurs between all nodes, and for time steps $40 \leq n < 70$, nodes 1 and 2 have no communication with node 3 (these time instants are indicated on the object's trajectory with cyan 'x's). For comparison, the performance of a centralized Kalman filter (using the same data) is also given.

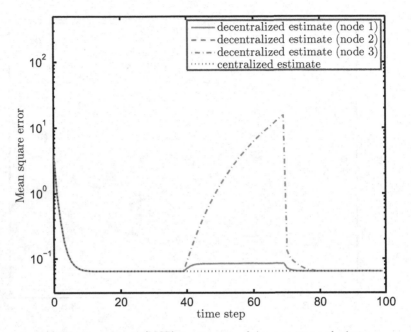

(c) Mean square error (MSE) in estimation of object position and velocity

Figure 4.4c: Distributed object tracking using the decentralized Kalman filter. The object moves according the dynamical law given in (3.12), with $\Delta t = 0.5$ and $\sigma_x = 0.1$. Noisy measurements are collected using $K = 3$ sensor nodes as specified in (4.16), with $\sigma_y = 0.25$. $N = 100$ time steps of decentralized Kalman filtering are performed and the actual and estimated object position and the mean square error in position and velocity are shown in the plots. For time steps $n < 40$ and $n \geq 70$, perfect communication occurs between all nodes, and for time steps $40 \leq n < 70$, nodes 1 and 2 have no communication with node 3 (these time instants are indicated on the object's trajectory with cyan '×'s). For comparison, the performance of a centralized Kalman filter (using the same data) is also given.

CHAPTER 5

Conclusion

In this manuscript, the Kalman filter was introduced as a tool for statistical estimation in the context of linear Gaussian dynamical systems. Several example applications of the Kalman filter were explored, along with extensions and implementation in MATLAB. Estimation techniques were first introduced in Chapter 2. Methods and example applications were presented for maximum-likelihood, Bayesian, and sequential Bayesian estimators. The Kalman filter was derived as the sequential Bayesian solution to an example problem of estimating the state of a dynamical system evolving according to a simple 1-D linear Gaussian state space model. In Chapter 3, the Kalman filter was presented formally as an estimation algorithm for the general multivariate linear Gaussian dynamical system. Several examples of Kalman filtering were given, including its relation to the RLS adaptive filter. Extensions of the Kalman filter were discussed in Chapter 4. The extended Kalman filter was outlined as a solution to the nonlinear Gaussian estimation problem. A framework for distributed estimation using Kalman filtering was also presented. In each case, examples and applications were provided. The examples were supplemented with MATLAB implementation, and graphs generated from these programs to better illustrate the concepts.

This manuscript serves as an introduction to Kalman filtering and provides a glimpse into some of its features, applications, and extensions. For further coverage interested readers are referred to the literature: for general Kalman filtering [6, 12, 35, 36], Kalman filtering and MATLAB [37], Kalman filtering implementation and applications [38–40], and Kalman filter extensions [5, 41].

Notation

$\mathbf{x}$	parameters of interest; state vector
D	dimension of state vector
$\mathbf{y}$	observed data; measurement vector
M	dimension of measurement vector
$p(\cdot)$	probability density function (pdf)
$p(\cdot\vert\cdot)$	conditional pdf
$\hat{\mathbf{x}}$	estimate of state $\mathbf{x}$
N	number of measurements; number of time steps
$E[\cdot]$	expected value
$\mathbf{I}$	identity matrix
$[\ldots]^T$	vector or matrix transpose
n	discrete time
r_i^{xy}	cross-correlation of random processes x_n and y_n for lag i
$\mathbf{x}_n$	state vector at time step n
$\mathbf{y}_n$	measurement vector at time step n
$\mathbf{Y}_n$	set of measurements up to time step n
$\hat{\mathbf{x}}_n$	estimate of state vector at time step n
$\mathbf{F}_n$	state-transition matrix at time step n
$\mathbf{v}_n$	state noise vector at time step n
$\mathbf{Q}_n$	state noise covariance matrix at time step n
$\mathbf{H}_n$	measurement matrix at time step n
$\mathbf{w}_n$	measurement noise vector at time step n
$\mathbf{R}_n$	measurement noise covariance matrix at time step n
$\mathcal{N}$	Gaussian distribution
m	mean
σ	standard deviation
$\mathbf{m}_{n\vert n}$	mean of Gaussian posterior state distribution at time step n computed using measurements up to time step n
$\mathbf{P}_{n\vert n}$	covariance of Gaussian posterior state distribution at time step n computed using measurements up to time step n
$\mathbf{f}(\cdot)$	state-transition function
$\mathbf{h}(\cdot)$	measurement function
$\tilde{\mathbf{F}}_n$	Jacobian of state-transition function evaluated at $\mathbf{m}_{n-1\vert n-1}$
$\tilde{\mathbf{H}}_n$	Jacobian of measurement function evaluated at $\mathbf{m}_{n\vert n-1}$

K	number of sensor nodes
$\mathbf{y}_{n,k}$	measurement vector at time step n for node k
M_k	dimension of measurement vector for node k
$\mathbf{H}_{n,k}$	measurement matrix at time step n for node k
$\mathbf{w}_{n,k}$	measurement noise vector at time step n for node k
$\mathbf{R}_{n,k}$	measurement noise covariance matrix at time step n for node k

Bibliography

[1] H. L. V. Trees, *Detection, Estimation, and Modulation Theory, Part I.* Wiley Interscience, 2001. DOI: 10.1002/0471221082.

[2] A. Papoulis and S. U. Pillai, *Probability, Random Variables and Stochastic Processes*, 4th ed. McGraw-Hill, 2002.

[3] R. Kalman, "A new approach to linear filtering and prediction problems," *Trans. ASME Ser. D. J. Basic Eng.*, vol. 82, pp. 35–45, 1960. DOI: 10.1115/1.3662552.

[4] R. Kalman and R. Bucy, "New results in linear filtering and prediction theory," *Trans. ASME Ser. D. J. Basic Eng.*, vol. 83, pp. 95–108, 1961. DOI: 10.1115/1.3658902.

[5] B. Ristic, S. Arulampalam, and N. Gordon, *Beyond the Kalman Filter: Particle Filters for Tracking Applications.* Artech House, 2004.

[6] Y. Bar-Shalom, X. R. Li, and T. Kirubarajan, *Estimation with Applications to Tracking and Navigation*, 1st ed. New York: John Wiley & Sons, 2001. DOI: 10.1002/0471221279.

[7] B. Cipra, "Engineers look to Kalman filtering for guidance," *SIAM News*, vol. 26, 1993.

[8] M. S. Grewal, L. R. Weill, and A. P. Andrews, *Global Positioning Systems, Inertial Navigation, and Integration.* Wiley-Interscience, 2001.

[9] D. Titterton and J. Weston, *Strapdown Inertial Navigation Technology*, 2nd ed. AIAA, 2004. DOI: 10.1049/PBRA017E.

[10] R. O. Duda, P. E. Hart, and D. G. Stork, *Pattern Classification*, 2nd ed. Wiley Interscience, 2001.

[11] D. J. C. MacKay, *Information Theory, Inference, and Learning Algorithms.* Cambridge University Press, 2003.

[12] S. Haykin, *Adaptive Filter Theory*, 4th ed. Prentice Hall, 2001.

[13] A. Gelman, J. B. Carlin, H. S. Stern, and D. B. Rubin, *Bayesian Data Analysis*, 2nd ed. CRC Press, 2004.

[14] T. Bayes and R. Price, "An essay towards solving a problem in the doctrine of chances. By the late rev. Mr. Bayes, F. R. S. Communicated by Mr. Price, in a letter to John Canton, A. M. F. R. S." *Philosophical Transactions*, vol. 53, pp. 370–418, 1763. DOI: 10.1098/rstl.1763.0053.

[15] W. R. Gilks, S. Richardson, and D. J. Spiegelhalter, Eds., *Markov Chain Monte Carlo in Practice*. Chapman & Hall/CRC, 1996. DOI: 10.1007/978-1-4899-4485-6.

[16] A. Doucet, N. D. Freitas, and N. Gordon, Eds., *Sequential Monte Carlo methods in practice*, 1st ed. Springer, 2001. DOI: 10.1007/978-1-4757-3437-9.

[17] G. H. Golub and C. F. V. Loan, *Matrix Computations*, 3rd ed. Baltimore: Johns Hopkins University Press, 1996.

[18] S. Edla, J. Zhang, J. Spanias, N. Kovvali, A. Papandreou-Suppappola, and C. Chakrabarti, "Adaptive parameter estimation of cardiovascular signals using sequential bayesian techniques," in *Signals, Systems and Computers (ASILOMAR), 2010 Conference Record of the Forty Fourth Asilomar Conference on*, nov. 2010, pp. 374 –378. DOI: 10.1109/ACSSC.2010.5757538.

[19] K. Tu, H. Thornburg, M. Fulmer, and A. Spanias, "Tracking the path shape qualities of human motion," in *Acoustics, Speech and Signal Processing, 2007. ICASSP 2007. IEEE International Conference on*, vol. 2, april 2007, pp. II–781 –II–784. DOI: 10.1109/ICASSP.2007.366352.

[20] R. E. Lawrence and H. Kaufman, "The Kalman filter for the equalization of a digital communications channel," *IEEE Transactions on Communication Technology*, vol. COM-19, pp. 1137–1141, 1971. DOI: 10.1109/TCOM.1971.1090786.

[21] A. B. Narasimhamurthy, M. K. Banavar, and C. Tepedelenlioglu, *OFDM Systems for Wireless Communications*. Morgan & Claypool Publishers, 2010. DOI: 10.2200/S00255ED1V01Y201002ASE005.

[22] N. Nair and A. Spanias, "Fast adaptive algorithms using eigenspace projections," in *Signals, Systems and Computers, 1994. 1994 Conference Record of the Twenty-Eighth Asilomar Conference on*, vol. 2, oct-2 nov 1994, pp. 1520 –1524 vol.2. DOI: 10.1109/ACSSC.1994.471712.

[23] N. Nair and A. Spanias, "Gradient eigenspace projections for adaptive filtering," in *Circuits and Systems, 1995., Proceedings., Proceedings of the 38th Midwest Symposium on*, vol. 1, aug 1995, pp. 259 –263 vol.1. DOI: 10.1109/MWSCAS.1995.504427.

[24] C. Panayiotou, A. Spanias, and K. Tsakalis, "Channel equalization using the g-probe," in *Circuits and Systems, 2004. ISCAS '04. Proceedings of the 2004 International Symposium on*, vol. 3, may 2004, pp. III – 501–4 Vol.3. DOI: 10.1109/ISCAS.2004.1328793.

[25] J. Foutz and A. Spanias, "An adaptive low rank algorithm for semispherical antenna arrays," in *Circuits and Systems, 2007. ISCAS 2007. IEEE International Symposium on*, may 2007, pp. 113 –116. DOI: 10.1109/ISCAS.2007.378234.

[26] T. Gupta, S. Suppappola, and A. Spanias, "Nonlinear acoustic echo control using an accelerometer," in *Acoustics, Speech and Signal Processing, 2009. ICASSP 2009. IEEE International Conference on*, april 2009, pp. 1313 –1316. DOI: 10.1109/ICASSP.2009.4959833.

[27] MATLAB documentation for the RLS adaptive filter function adaptfilt.rls, available online at http://www.mathworks.com/help/toolbox/dsp/ref/adaptfilt.rls.html.

[28] R. Olfati-Saber, "Distributed Kalman filtering for sensor networks," in *46th IEEE Conference on Decision and Control*, New Orleans, LA, 2007, pp. 5492–5498. DOI: 10.1109/CDC.2007.4434303.

[29] C. Y. Chong, S. Mori, and K. C. Chang, "Distributed multitarget multisensor tracking," in *Multitarget-Multisensor Tracking: Advanced Applications*, Y. Bar-Shalom, Ed.   Artech House, 1990, ch. 8, pp. 247–295.

[30] K. C. Chang, C. Y. Chong, and Y. Bar-Shalom, "Distributed estimation in distributed sensor networks," in *Large-Scale Stochastic Systems Detection, Estimation, Stability and Control*, S. G. Tzafestas and K. Watanabe, Eds.   Marcel Dekker, 1992, ch. 2, pp. 23–71.

[31] S. J. Julier and J. K. Uhlmann, "A new extension of the Kalman filter to nonlinear systems," in *Proceedings of AeroSense: The 11th Symposium on Aerospace/Defence Sensing, Simulation and Controls*, Orlando, FL, 1997.

[32] A. A. Berryman, "The orgins and evolution of predator-prey theory," *Ecology*, vol. 73, pp. 1530–1535, 1992. DOI: 10.2307/1940005.

[33] C. Moler, "Experiments with MATLAB," http://www.mathworks.com/moler/exm/index.html, 2011.

[34] B. S. Y. Rao, H. F. Durrant-Whyte, and J. A. Sheen, "A fully decentralized multi-sensor system for tracking and surveillance," *The International Journal of Robotics Research*, vol. 12, pp. 20–44, 1993. DOI: 10.1177/027836499301200102.

[35] T. Kailath, A. H. Sayed, and B. Hassibi, *Linear Estimation*.   Prentice Hall, 2000.

[36] D. Simon, *Optimal State Estimation: Kalman, H Infty, and Nonlinear Approaches*.   John Wiley and Sons, 2006. DOI: 10.1002/0470045345.

[37] M. S. Grewal and A. P. Andrews, *Kalman Filtering: Theory and Practice Using MATLAB*. John Wiley and Sons, 2008. DOI: 10.1002/9780470377819.

[38] R. L. Eubank, *A Kalman Filter Primer*.   CRC Press, 2006.

[39] C. K. Chui and G. Chen, *Kalman Filtering: with Real-Time Applications*.   Springer, 2009.

[40] P. Zarchan and H. Musoff, *Fundamentals of Kalman Filtering: A Practical Approach*, 3rd ed. American Institute of Aeronautics and Astronautics, Inc., 2009.

[41] S. Haykin, Ed., *Kalman Filtering and Neural Networks*.   John Wiley and Sons, 2001. DOI: 10.1002/0471221546.

Authors' Biographies

NARAYAN KOVVALI

Narayan Kovvali received the B.Tech. degree in electrical engineering from the Indian Institute of Technology, Kharagpur, India, in 2000, and the M.S. and Ph.D. degrees in electrical engineering from Duke University, Durham, North Carolina, in 2002 and 2005, respectively. In 2006, he joined the Department of Electrical Engineering at Arizona State University, Tempe, Arizona, as Assistant Research Scientist. He currently holds the position of Assistant Research Professor in the School of Electrical, Computer, and Energy Engineering at Arizona State University. His research interests include statistical signal processing, detection, estimation, stochastic filtering and tracking, Bayesian data analysis, multi-sensor data fusion, Monte Carlo methods, and scientific computing. Dr. Kovvali is a Senior Member of the IEEE.

MAHESH BANAVAR

Mahesh Banavar is a post-doctoral researcher in the School of Electrical, Computer and Energy Engineering at Arizona State University. He received the B.E. degree in Telecommunications Engineering from Visvesvaraya Technological University, Karnataka, India, in 2005, and the M.S. and Ph.D. degrees in Electrical Engineering from Arizona State University in 2007 and 2010, respectively. His research area is Signal Processing and Communications, and he is specifically working on Wireless Communications and Sensor Networks. He is a member of MENSA and the Eta Kappa Nu honor society.

ANDREAS SPANIAS

Andreas Spanias is Professor in the School of Electrical, Computer, and Energy Engineering at Arizona State University (ASU). He is also the founder and director of the SenSIP Industry Consortium. His research interests are in the areas of adaptive signal processing, speech processing, and audio sensing. He and his student team developed the computer simulation software Java-DSP (J-DSP–ISBN 0-9724984-0-0). He is author of two text books: *Audio Processing and Coding* by Wiley and *DSP; An Interactive Approach*. He served as Associate Editor of the *IEEE Transactions on Signal Processing* and as General Co-chair of IEEE ICASSP-99. He also served as the IEEE Signal Processing Vice-President for Conferences. Andreas Spanias is co-recipient of the 2002 IEEE Donald G. Fink paper prize award and was elected Fellow of the IEEE in 2003. He served as Distinguished Lecturer for the IEEE Signal Processing Society in 2004.

Printed in the United States
by Baker & Taylor Publisher Services